# Perfect Planet, Clever Species

## HOW UNIQUE ARE WE?

**WILLIAM C. BURGER**
*Curator Emeritus, Field Museum of Natural History*

Prometheus Books
59 John Glenn Drive
Amherst, New York 14228-2197

Published 2003 by Prometheus Books



Inquiries should be addressed to
Prometheus Books
59 John Glenn Drive
Amherst, New York 14228–2197
VOICE: 716–691–0133, ext. 207
FAX: 716–564–2711
WWW.PROMETHEUSBOOKS.COM

07 06 05 04 03 5 4 3 2 1

Library of Congress Cataloging-in-Publication Data

Burger, William C.
Perfect planet, clever species : how unique are we? / William C. Burger.
p. cm.
Includes bibliographical references and index.
ISBN 1–59102–016–6 (alk. paper)
1. Human evolution. 2. Life on other planets. 3. Exobiology. I. Title.

GN281.4 .B87 2002
599.93'8—dc21

2002068117

Printed in Canada on acid-free paper

perfect planet.

# CONTENTS

clever species

perfect planet. clever species

# ACKNOWLEDGMENTS

My home institution, the Field Museum, has played a central role in the development of this book. Supporting four major scientific disciplines, the museum's libraries include publications in anthropology, botany, geology, and zoology. This has given me the opportunity to scan the incoming journals of these various fields, and interact with a great variety of scholars, over more than thirty years. In addition, symposia and seminary series, both here at the Field Museum and at the University of Chicago, have been very helpful in learning about new ideas bubbling up in the natural sciences. However, as a scientific institution focused on original research, the museum's libraries hold few titles outside its area of specialization. This is where Chicago's central public library has played a crucial role. Housing a variety of introductory, popular, and basic technical books accessible on open stacks, the Harold Washington Library has been essential in surveying a broad literature. In addition, journals such as *Nature*, *Science*, *Scientific American*, *Bioscience*, and others have made it possible to keep up with recent advances across a wide spectrum of disciplines. Finally, there are the authors of the many books cited among the references, both journalists and scientists, who have labored to make an ever-expanding mountain of scientific study comprehensive to the general reader. Their books have been very useful in crafting this volume.

In addition, many people I have worked with over many years have contributed to the ideas presented in this book, both directly and indirectly. They are too numerous to name, but I would like to thank Matthew Nitecki and Edward Yastrow, who read early versions of the manuscript and made significant suggestions. John Albright, Deborah Bakken, Barbara Becker, Cheryl Braunstein, Bennet Bronson, and Jack Fooden read parts of the manuscript and also made helpful criticisms. And then there are the other Burgers—Melinda, Helen, and Carolyn—who have listened to these ideas over many years, and whose responses helped shape many of the arguments. But creating a good book requires other kinds of help as well, especially a sharp editor. Linda Regan's efforts resulted in a shorter, better-phrased, and more reader-friendly text.

Unfortunately, when covering so many fields of inquiry, there are many opportunities to pick up false impressions or to misinterpret research reports. I have tried to avoid such errors, but I'm sure a number remain. Not a few readers will disagree strongly with some of the views promoted here. In the case of the more important issues, I have tried to cite alternative opinions, either in the text or in the notes. But whatever views others might hold, I believe we can all agree that we live on a glorious planet, and that our intellectual achievements have been quite amazing.

William C. Burger
8 September 2002

perfect planet.

# INTRODUCTION

clever species

As we begin the twenty-first century, we mark the end of one of the most significant millennia in human history. This was the period during which Western Europe combined dynamic mercantile interchange, technological creativity, and an unquenchable thirst for knowledge to forge the technological culture that has come to rule our planet. The scientific and industrial revolutions of the last millennium have transformed human potential and changed the surface of our planet. These dramatic changes were the product of an even more profound revolution, ten thousand years earlier.

Shortly after the ice of the last glacial episode had retreated, our species began a transformation; the archaeological record of this period bears witness to a major cultural innovation. Beginning as early as ten millennia ago, and in several different parts of the world, we humans selected a small number of plant and animal species to become closely integrated partners in our daily lives. That symbiotic transformation, the agricultural revolution, provided human communities with a more generous and reliable food supply. Historians may complain that this bold advance also multiplied pestilence and warfare, but it mattered little. People in agricultural communities became much more numerous than their hunting and gathering antecedents. About five millennia later, and in only a very few locations, farming communities would come together to create larger and more complex city-states. Within this new cultural setting metallurgy developed, writing emerged, and metaphysics blossomed. Over the next few thousand years, civilizations would rise and fall until the seeds of another cultural revolution took root.

Five hundred years after the fall of Rome, at the beginning of the second millennium of the Christian calendar, western Europe experienced better weather. Harvests grew more bountiful, forests gave way to plowed fields, populations expanded, and with them towns and commerce flourished. The translation of Greek and Arabic scholarship into Latin came at a time when Christian philosophers were exploring the possibility of appreciating what they saw as God's majesty through the study of nature. Also, as ecclesiastical law became separated from secular law, Europe was transformed into a more open society. Then came paper, the printing press, the discovery of a much larger world, and the Protestant Reformation. These fortunate coincidences created the environment from which the scientific revolution would emerge. Five hundred years later, using the methodologies of science and technology (and a lot of fossil fuel) we have, for better and for worse, changed our relationship with the planet that sustains us.

Now that we are living within a dynamic scientific-industrial civilization, some people are convinced we're experiencing visits by beings from more advanced civilizations elsewhere in the universe. Earlier human communities believed in spirits, leprechauns, and ghosts; now we've added mysterious aliens from outer space. So far there is no convincing evidence for such visits. But how likely might such encounters be? How rare or common might our own odyssey on this planet be? Are we alone?

By reviewing what science has been able to piece together regarding our planet, our star, and our history, these pages attempt to help set the stage for addressing such provocative questions. Over the last half century, science has discovered the nature of the genetic code and we are now beginning to understand how this code builds proteins, controls development, and creates the diversity of life adorning our planet. Space probes have revolutionized our understanding of the solar system as geologists have come to realize how dynamic a planet the Earth really is. The variety of life today and its history over time is being documented in ever greater detail and depth.

It is easy to lose a broader perspective in the midst of an explosion of new scientific information. Scanning this new knowledge, the following chapters attempt to summarize the history of our planet and our lineage. Our story has many developments that we have begun to understand, but there have also been important historical accidents. Reviewing this history should help us address the question of the likelihood that other technological civilizations share our galaxy.

In some quarters the idea of organic evolution continues to be challenged or qualified as an "unproven theory." One can counter this claim with volumes of correlated data sets, but one of the most obvious and dramatic pieces of evidence for evolution has nothing to do with biology or human origins at all—it is the Atlantic Ocean. For half a century American geologists refused to even consider the idea that the continents on either side of the Atlantic had come asunder and "drifted" apart. Yet any child can look at a globe and see the obvious symmetry of the Atlantic's eastern and western shores. Finally, after decades of intensive study, geologists had to admit that the continents had indeed come asunder, and that the Atlantic has expanded over time. In a few localities, geological strata on either side of the Atlantic confirm the fact that once, long ago, there was no Atlantic Ocean. Just as the Atlantic has "evolved," so has just about everything else on this planet.[1]

Careful studies of isotopic ratios of elements in rocks fallen from the sky (meteorites), analyses of fossils deeply imbedded in the earth, and comparisons of the genetic code of diverse lineages of plants and animals, all tell the same concordant story. Our planet and its life-forms developed over thousands of millions of years. We have evolved, oceans have evolved, and even the Earth's atmosphere has evolved. Thanks to the activities of the living world, we now have an atmosphere rich in oxygen, poor in carbon dioxide, and made vibrant by birds, bats, and butterflies. Evolution is a fact. Only its tempo, direction, and meaning are contentious.

Science has been a powerfully successful enterprise in our quest to understand nature. Its goal is to interpret natural processes in terms of natural forces. Scientific questions are ultimately tied to physical realities that others can witness or verify for themselves. Science requires observational or experimental data, careful objective analyses capable of repetition and confirmation, hypotheses consistent with the data, and testable theories capable of revision or rejection. Albert Einstein suggested that the gravitational field of the Sun would bend the weightless light of a distant star. This energized the physics community, which eagerly awaited images of stars photographed close to the Sun's disc during a total solar eclipse. These images were carefully compared with photographs made earlier when the light of those same stars was far from the Sun. Comparisons of the photographs showed that the stars' images had indeed been shifted when passing close to the Sun—almost exactly as Einstein had predicted. In contrast, Charles Darwin's and Alfred R. Wallace's explanation for evolution by natural selection is a less precise theory and not so easily verifiable. Nevertheless, the theory of natural selection has spawned innumerable subsidiary hypotheses whose confirmation has given the general theory an overwhelming sense of validation.[2]

Despite many triumphs, scientific studies rarely engage the imagination of the majority of humankind. Today, as in the past, a great many people find that traditional religious beliefs have profound meaning for their own daily lives, and most are convinced that we humans and our origin stand apart from the rest of the natural world. Such beliefs are a very different aspect of our intellectual life, addressing many issues beyond the scope of scientific inquiry. The belief that we and the universe were created according to the plan of an omnipotent and purposeful deity is central to many religious and philosophic traditions. Though deeply reassuring, such beliefs do not permit the formulation of testable hypotheses. In such a providential universe, humankind might be unique, or there might be billions of others just like us, scattered through the firmament. There is no way to guess what the deity has done elsewhere in the universe. Providential creation and divine direction are ideas that cannot be further tested, and for that reason they are excluded from the playing fields of science.

The scientific approach, with or without a divine presence, assumes that we have come into being and our planet has evolved subject to the forces and contingencies we can see and analyze around us, or infer from ancient rocks and the light of distant stars. The scientific approach has the advantage of allowing us to set forth alternative hypothetical scenarios that are amenable to critical analysis. Today, we can supplement studies of fossil organisms with careful analyses of the genetic codes of their living descendants. The complex patterns in the genetic code mix ancient instructions with modern variations, duplications, and innovations. Using DNA sequences, we have been able to tease out historical patterns in the genes and gain additional insights into the history of life. Almost without exception, the results from the DNA analyses concur with the fossil record, and it appears that we are indeed unraveling ancient mysteries in a realistic fashion. In all these cases, we have been looking for data sets that are verifiable through repeated observation or analysis, a fundamental aspect of all scientific inquiry. New data, contradicting previous assumptions or illuminating deeper generalities, have been critically important to scientific progress. Experimental results or new disclosures that can falsify at least one of several proposed explanations are especially valuable. All such scientific activities assume that we can understand the processes and forces of nature without resorting to unverifiable mystical or providential explanations. Five centuries of repetitive observations with ever more critical experimentation and observation have brought scientific knowledge ever closer to the realities of nature.

This body of knowledge has given rise to a huge, internally consistent historicism with which we attempt to understand our universe and ourselves. Woven together from many loosely independent subject disciplines, modern science has created a grand historical narrative that informs these pages. The ecologist Robert MacArthur once suggested that biologists come in two varieties: engineers and historians. The engineers want to know how things work while the historians want to figure out how they got that way. We'll be mostly historians as we traverse four billion years of life on Earth. But we'll also need to look at how things work and interact to help us understand this long epic more fully.

In October 1993 the United States Congress ended federal funding for radio telescope projects committed to the Search for Extraterrestrial Intelligence (SETI). American taxpayers had been spending up to ten million dollars a year on this enterprise. Focused on those frequencies in the radio spectrum that had the clearest "window" for long-distance communication, giant radio telescopes had logged thousands of hours of careful listening. Though no longer funded by the federal government, the search for advanced extraterrestrial civilizations continues using a variety of technologies through private funding.

Those who feel that advanced technological societies are likely to be common in our galaxy have included some astronomers and physicists, appreciative of the grand numbers of stars that populate the Milky Way. Biologists, however, seem to be far more pessimistic, perhaps because of an awareness of the many historical contingencies that seem to have played so important a role in our long history on this lovely planet. The reasons for such a view are based not only on a lot of "lucky breaks" in the history of life on Earth, but our planet itself appears to have experienced a number of fortuitous events early in its history. On the other hand, several aspects of humankind's long odyssey appear to be the product of strong directional trends. These pages attempt to traverse the full span of Earth's long history, focusing on those key events and developments that have put us where we are today. The scientific endeavors of the last five hundred years have disclosed a grand and richly detailed history. Using that history, the following chapters examine the critical episodes of our planet's early history and the long trajectory of life, as well as recent cultural advances. Recounting the highlights of this story should place the question of extraterrestrial intelligence in sharper focus and give us a better notion of how likely it is that we will be communicating with other technological civilizations and their radio telescopes elsewhere in our galaxy. Let us start at the beginning of our odyssey, with the origin and nature of the object that energizes our every breath, the Sun.

# I

# NO ORDINARY STAR

clever species

The notion that there might be thousands of advanced civilizations in our galaxy is based on many assumptions, the most fundamental of which takes the Sun to be an ordinary and common star. With perhaps as many as 300,000 million stars in the Milky Way galaxy, an assumption that stars are more or less all alike makes other technological civilizations seem just about inevitable.[1] (Since we'll be discussing many historical scenarios over these huge time periods, I think it's better to use "thousand million" as a basic measure, avoiding the word *billion*. Using the same basic measure of a million will make it easier to compare early episodes in the history of our planet with those that took place more recently.) But our Sun is not an average star, and our galaxy is not just a sack of randomly assorted similar stars.

The Sun, our star, is located in an immense, flattened, disk-shaped aggregation of stars we call the Milky Way. The Milky Way galaxy is, in turn, part of the "local cluster," a group of about thirty galaxies. Most of these galaxies are small, like the Magellanic Cloud of the Southern Hemisphere's night sky, but two are huge. The giant spiral galaxy in the constellation of Andromeda is the largest, with our own galaxy coming in a close second, at about 130,000 light-years wide. (Traversing almost six trillion miles in a year, the speed of light is a universal constant, which travels 186,282 miles or 300,000 km *in a second.*) Our galaxy has the shape of a pancake with a bulging center about 10,000 light-years thick where stars are numerous and closely congested. The thin, broad, circular disk is only about 5,000 light-years thick and includes aggregations of stars and dust in "arms" spiraling outward from the center. Surrounding the visible disk is a dark, much larger, and very sparsely populated spherical region of stars designated the "halo." Many stars in the halo are ancient, probably formed in the early history of our galaxy, before it had spun its way into the form of a flattened spiral. Today, the youngest stars and active star-forming regions are largely confined to the thin disk where interstellar gases and dust have been concentrated along the galaxy's central plane. Our spiral galaxy is dynamic, slowly rotating in grandly cadenced synchrony over hundreds of millions of years. However, there's a big problem: the visible stars don't seem to provide nearly enough mass to keep everything from flying apart. Astronomers have met this challenge by postulating the existence of invisible "cold dark matter" to hold things together. Or it may be that there are aspects of our universe that science is not yet aware of—and we really don't know as much as we think we do.[2]

Positioned about halfway from the center of the galaxy in the more densely populated central layer of the disk, the Sun appears to differ little from a great many other stars. However, as we shall see, our star is neither ordinary nor average. From the point of view of classification by color and luminosity, our Sun fits comfortably within a grouping of common stars called the "main sequence." These stars derive their powerful radiant energy from the conversion of a small amount of mass which is lost as hydrogen nuclei fuse to form helium. This process takes place under tremendous pressure and million-degree temperatures, deep within the star. The amount of mass "lost" during the conversion of hydrogen atoms to helium atoms may be small, but the result is a huge amount of energy, now understood thanks to Einstein's fundamental equation: $E = MC^2$. This tells us that the energy yield is the mass multiplied by the speed of light multiplied by itself—a humongous factor. Every second, the Sun converts about 5 million tons of mass into energy. Small and distant, our planet intercepts only a billionth of this continuously outflowing bounty.

Astronomers calculate that the very large stars in the main sequence burn their hydrogen at such a rapid rate that they may run out of fuel after only a few hundred million years. In the case of Sirius, with twice the mass of our sun and twenty-five times the luminosity, astronomers estimate a life span of only 1,000 million years. Smaller stars, those with a mass about 1.5 times that of our Sun, are calculated to have a life span of about 3,500 million years. If our Sun had such a life span, we simply wouldn't be here. About 4,000 million years passed before complex animals made their first appearance here on planet Earth. Not only do larger stars have a shorter life expectancy, they are quite rare and emit high levels of life-threatening ultraviolet radiation—not a friendly environment for living things that aspire to live on land.

Based on age estimates of the oldest meteorites, our Sun probably originated around 4,500 million years ago and is now near the midpoint of its life span.[3] Stars equaling or smaller than the Sun are calculated to have life spans in excess of 10,000 million years, affording lots of time in which to evolve life and elaborate complex ecosystems.

As is true for so many other classes of objects, it is the smallest stars that are the most common. In a fine discussion of the universe, Ken Croswell estimates that 95 percent of our galaxy's stars are smaller than the Sun.[4] (The majority of these are called class M red dwarf stars.) Because they burn their fuel more slowly, small stars have even longer lives than the Sun—but they radiate less energy. Despite their being common, we do not see these stars when we look out into a starry night; they are simply too dim. Through the telescope they appear reddish because they radiate most of their energy in the red and infrared regions of the spectrum. Those longer wavelengths carry much less energy than the shorter yellow, blue, and ultraviolet wavelengths. This means that the orbital distance within which a planet might maintain liquid water on its surface must be much closer to the star, and this "planetary comfort zone" will be confined within a narrower range as well. The likelihood of having an Earthlike planet in exactly that narrow comfort zone around a smaller class M dwarf star is small. Being closer to the parent star presents an additional and serious problem; the gravitational force of the star may come to "lock" the planet's revolution so that one side always faces the star, or slows the planet's rotation down to produce extremely long "days." Mercury's period of rotation (its day) is fifty-eight Earth days long in a year of eighty-eight Earth days. Such long days on a slowly revolving planet can produce a furnacelike surface on one side while the other side freezes.

Another problem with the smaller and common class M red dwarf stars is that they are not especially stable, occasionally producing flares that could wipe out terrestrial life-forms. All told, class M stars, comprising about 80 percent of the stars in our galaxy, are utterly unlikely to have life-inhabiting planets. Of all the stars, Croswell suggests that only the cooler F stars, all the Sunlike G stars, and the warmer K stars are likely to be suitable for the evolution of life. Not too big, not too small, and comfortably stable over the long haul, it appears that we have got ourselves just the right kind of star.

## LOCATION, LOCATION, LOCATION

Our splendid star has provided a stable energy source for an extraordinary proliferation of living things over a very long period. Such an extended history required not only a stable star, but also a safe environment. We might recall the claim made for choosing real estate: "The three most important criteria are location, location, location." Since a star like the Sun will orbit the center of the galaxy many times over the thousands of millions of years required for the evolution of complex life-forms, the first choice of location for the evolution of life should be a quiet region of the galaxy. This location must be in a region where deadly radiation or frequent supernova explosions are unlikely to be encountered over the star's long journey. The center of the galaxy has regions where bursts of intense X-ray and ultraviolet radiation have been very common—not a good neighborhood for complex living things sensitive to high-energy radiation.

Situated about 28,000 light-years from the center, our star is about halfway between the hyperactive middle and the outer reaches of the galaxy. As part of the peripheral galactic disk, the Sun is located within the innermost "thin disk" (about 1,000 light-years thick) which is, in turn, sandwiched between the two outer layers of the "thick disk." As we look directly toward the center of the galaxy on a clear night, we see the Milky Way as a narrow, luminous band across part of the sky. In all other directions we see the stars of the disk within which we ourselves are traveling. Together with most other stars of our region, our Sun sails along at a speed of about 26,800 miles per hour (12 km/sec), circling the galactic center every 250 million years, in cadence with the majestically rotating spiral arms. As it travels, the Sun oscillates a bit, moving up and down across the galactic plane every 33 million years. Astronomer Priscilla Frisch reports that, on a local scale, we have been traveling through a particularly dust-free region over the last few million years, and she "wonders if it is a coincidence that *Homo sapiens* appeared while the Sun was traveling through a region of space virtually devoid of interstellar matter."[5] All told, the Sun meets the first criterion for a good location: a quiet neighborhood.

The second critical location requirement for the creation of life is that the star is formed in a region where sufficient heavier elements were available to form sizeable rocky planets and to eventually construct complex organic molecules. With hydrogen and helium making up more than 98 percent of the material of our galaxy, astronomers use the term "metallicity" for the abundance of all elements heavier than helium. Heavier elements contain more protons and neutrons in their nuclei, but they are *not* distributed evenly throughout the galaxy. They are common in the center of the galaxy where dying stars have produced a rich mix of elements. Heavier elements are extremely rare among the older stars of the outer halo, and they are not particularly common in the spiral arms of the galactic disk. To beget life, a star must be born from within a dust cloud fertilized with an ample supply of those heavier elements central to the lives of complex living things. Fulfilling our second criterion parameter, our Sun is such a star—rich in metallicity and graced with a flock of rocky planets and icy comets.

Finally, for life to evolve, a star must have an appropriate solid planet revolving at just the right orbital distance to keep water wet and life comfortable. This is the third critical location parameter. A planetary orbit too far from its star will keep water permanently frozen. Too close, and the surface becomes a frying pan. The distance that can maintain water in its liquid state over time is the only likely location for the origin and survival of complex living things. Clearly, from a real-estate perspective, our Sun scored perfectly in regard to all three location parameters. But let's look at the galactic picture a little more closely.

Spectral analyses indicate that the Magellanic Cloud, a small satellite galaxy of the Milky Way, is deficient in heavy metals. Even the youngest stars in the Magellanic Cloud are poor in metals. Perhaps there was a paucity of supernova explosions in the history of our nearby galactic companion. It is only as a result of having many massive stars live and die that our own galaxy has acquired greater concentrations of heavier elements—but those heavier elements are not evenly distributed. The 500 million stars of the outer halo are estimated to be about twice as old as our Sun, but few have more than 10 percent of the metallicity of our star. Such percentages are not very likely to provide an appropriate environment for the evolution of the kind of life with which we are familiar.

Taking a broader perspective, astronomers Andrew and David Clark point out that our galaxy has, in all probability, a "habitation doughnut" where life is most likely to develop. The outer periphery beyond the doughnut is poor in the heavier elements essential to building complex life-forms; while the hole of the doughnut is the center of the galaxy where intense radiation and stellar explosions create a dangerous environment.[6] While a "pancake with a hole in the middle" might be a more accurate image, it is clear that our solar system, halfway out from the center of the galaxy, is located within the most likely region to support the development of rocky planets and complex life-forms.

## EXPLODED STARDUST

Physicists theorize that the heavier elements have been formed only under extremely high temperatures and pressures as huge stars collapse and then explode in a violent death (called a supernova), or in the last stages of the life of smaller stars. These ideas are imbedded in the big bang scenario of the birth and evolution of the universe. They imply that mass in the early universe was made up almost entirely of hydrogen, helium, and a little bit of lithium—early products of the original big bang itself. Later, helium "burning" in the nuclear furnaces of ordinary dying stars produced carbon and oxygen. In turn, the burning of hydrogen, oxygen, or carbon and their products produced neon, nitrogen, magnesium, silicon, sulfur, and iron. The creation of elements heavier than iron, however, is more difficult. Requiring extraordinarily high energies and pressures to forge their more complex heavy nuclei, such elements can be created only within the interior of a larger star in its final contraction and explosive ending. All together, theories of nucleosynthesis indicate that the heavier elements could have come into being only *after* an earlier generation of large, short-lived, explosive stars and smaller stars had gone through their life cycles. Cosmologists currently estimate the big bang origin of our universe to have taken place somewhere between 12,000 and 15,000 million years ago, plenty of time to produce the heavier elements essential for the formation of rocky planets and living things.[7]

In our particular corner of the Milky Way, the average metallicity of nearby stars appears to be about two-thirds that of our Sun. Specific evidence for our Sun's being unusual is the fact that it contains significantly more carbon than is found in similar stars in our region of the galaxy. Average abundance of carbon in nearby stars is estimated to be about 225 carbon atoms per million hydrogen atoms. Estimates for our favorite star range from 350 to 470 carbon atoms per million hydrogen atoms.[8] These figures strongly suggest that our Sun is "different" from most stars in the galaxy, having originated in a region especially rich in heavier elements. Luckily for us this region was richer not only in the common carbon atom but also in the rarer and heavier elements such as copper, zinc, tin, and many others. In fact, it may have been the shock wave of a nearby supernova explosion itself that produced a momentary compression within the presolar cloud. This may have both initiated the formation of our star and provided additional heavy elements that are so useful in building complex living things.

Meteorites, rocks that literally *have* fallen from the sky, are the Rosetta stones of the early solar system. Their maximum age, based on analyses of the ratios of radioactive elements and the decay products they contain, is calculated to be about 4,560 million years. Our planet was likely formed 10 to 100 million years later. Radioactive elements and unstable isotopes have become critical yardsticks in our measurement of time, ranging from the early days of the solar system to recent times. The isotopes of an element differ from each other by having more or fewer neutrons in their nuclei, slightly changing the atom's weight but not its chemical properties. More important, because isotopic differences lie deep within the nucleus, their characteristic weight cannot be altered by ordinary heat, pressure, or chemical reactions. Isotopes are therefore unaffected by even intensely stressful environmental conditions. However, some isotopes and a few elements are inherently unstable, and they decay to more stable forms over time, as measured by their half-life. The half-life of unstable elements and isotopes ranges from a few years to thousands of millions of years; it is simply the time it takes for half a population of these atoms to decay to the stable form. Again, because these changes occur deep within the atom's nucleus, environmental and chemical conditions do not change these half-lives. For example, uranium 238 loses mass and energy through thirteen intermediate steps until it finally becomes stable lead 206 over a half-life of 4,500 million years; the more common uranium 235 decays to lead 207 over a half-life of 704 million years. Once solidified into molten rock that has cooled and is unable to separate, the unstable atoms continue their clocklike decay, making dating possible by comparing the percentages of unstable atoms and their stable derivatives at later points in time. Unfortunately, radioactive elements frozen in crystallized rocks are rare, and there is always the problem of leakage or contamination. Ideally, more than one radioactive series should be compared from the same rock sample. But let's get back to rocks that have fallen from the sky.

Some meteorites apparently represent very early undifferentiated material, while others are the smashed remnants of larger asteroids that had become differentiated into a metallic central core with a rocky outer crust. In addition, a very few of the thousands of known meteorites have proven to be much younger than the rest and contain more hydrated minerals, suggesting that they came from a parent body with a longer independent history. Careful analyses have now convincingly shown that these rarely encountered and peculiar meteorites are rocks from Mars and the Moon. Careful analysis of the elemental composition of meteorites and their age distinguishes between the very few that are pieces of either Mars or the Moon, and the thousands that have come from more distant regions. During their formation and development, some of these early celestial bodies accumulated dustlike material from the original stellar cloud that gave rise to our solar system. Some of these minute grains have shown greatly enriched abundances of particular isotopes, which indicates they originated from unstable precursors formed within an exploding giant star.[9]

A cornucopia of heavier elements has played an important role in the history of the Earth. Many of the heavier elements are unstable, undergoing radioactive decay to lower-energy and more stable states, giving off a burst of energy as they make this transition. This is the source of energy that has kept the interior of our planet hot. In the Earth's early history, gravitational compaction and meteoritic impact also contributed heat. These effects, together with a generous dosage of radioactive elements (both long-lived and short-lived), made the Earth's interior really hot. This fostered Earth's internal *differentiation*, concentrating iron and other heavy elements in the core, while separating crust and mantle. Heat from radioactive decay continues to heat the Earth's core and contributes to a dynamic surface. But there's more to the importance of hefty elements.

A number of these heavier elements play an essential role in some of life's most critical biochemical reactions. Without iron, copper, magnesium, manganese, zinc, and other heavy metals, photosynthesis, respiration, and many other enzymatic processes might not have been possible. Copper and iron are at the center of enzymes that "burn" the oxygen we breathe; both magnesium and manganese play vital roles in enzymes that are central to photosynthesis. Zinc is necessary for enzymes that are part of the process of cell division, and cobalt is required for vitamin $B_{12}$. Many essential life processes cannot proceed unless they are catalyzed by enzymes configured around heavy atoms, elements that are quite literally bits of exploded stardust. Without such stardust it is unlikely that complex life forms could have developed on our planet.

In effect, the debris of dying stars and nearby supernova explosions enriched our solar system with the heavier elements that are so rare in the universe. Burning, dying, and exploding over thousands of millions of years, such stars have created a universe richer in elemental diversity over time. Consequently, it is only in an "older" universe that complex life forms can come into being. Based on the analysis of minute grains found associated with meteorites, Armand Delsemme suggests that our solar system was enriched by "at least four different stars. A star rich in carbon. A star rich in oxygen, a star rich in magnesium and silicon, and a star rich in iron."[10] Clearly, our glorious star was generously endowed with the materials needed to build living things. Together with a liquid medium such as water, complex life-forms built largely by the common and versatile carbon atom also require a variety of heavy elements to act as the energy centers of their most critical enzymes. From a distance, our star may look like millions of other stars, but present evidence suggests that it is far richer than most. Dame Fortune smiled early on our odyssey, when our star was graced with a wide assortment of heavier elements, amply fertilizing its little fleet of planets. Humankind is the product of that history; hydrogen, the original stuff of the universe, makes up about 8 percent of our body weight—all the rest of us is stardust.

## A VERY SPECIAL STAR

Besides its metallicity, there are other reasons to consider the Sun unusual. Admittedly, the Sun's size and spectral characteristics place it in the midst of the main sequence, an aggregation of stars including most of the stars in our galaxy and, probably, in most other galaxies as well. However, the Sun belongs to the warmer and more yellowish G class of stars that make up only about 5 percent of the stars in our galaxy. Their size, luminosity, and long stable lives make such stars the ideal hosts for a life-supporting planet such as Earth. While there may be minor cycles in these stars, such as the eleven-year sunspot cycle, they appear to be remarkably stable, radiating a nearly constant stream of energy into their surroundings over thousands of millions of years. It is this steady flow of sunshine that has warmed our planet and, with photosynthesis, fueled the evolution and proliferation of complex life-forms. Every time we think a thought or flex a muscle we are using the energy of sunlight—first captured by a green plant. Although some bacteria can live from the energy of simple chemical reactions and a few strange creatures may feed off bacteria living in the blackness near deep-ocean volcanic vents, our Sun is the primary energy source for 99 percent of the living activity on planet Earth.[11]

Not only does the Sun belong to an uncommon class of stars, it is unusual within this group as well: our Sun is a solitary star. A recent survey of Sunlike stars concluded that over 60 percent of such stars are not solitary at all, but members of binary pairs or triplets. An earlier survey of 164 G class Sunlike stars, thought to be representative of our region in the galaxy, found that only one-third were solitary stars. Alpha Centauri, our nearest celestial neighbor, is a triple-star system. Such stars are not the likely home for other civilizations. First of all, it seems probable that multiple-star systems would have swept up and absorbed all nearby material early in their formation and left little for the formation of planets. Second, the probability of stable planetary orbits of low eccentricity in regions receiving just enough energy to support life within a two-star or three-star system is vanishingly small. All told, when we imagine likely locales for the origin of complex life-forms, there seems to be no alternative to a stable, long-lived, metal-enriched, and *solitary* star.

Astronomer Rudolf Kippenhahn suggested that star formation may take two different paths. In one, most of the matter retains its angular momentum, forms a ringlike phase, and develops into a paired-binary star system. The other route, because of early separation of mass and angular momentum, produces a heavier, slowly rotating star in the center, while the planets retain nearly all the angular momentum of the system.[12] If these ideas are correct, nearly all solitary stars should have a retinue of planets, not unlike our own solar system.

The fact that a majority of Sunlike stars exist as binary pairs is also consistent with the idea that the giant gaseous outer planets of our solar system are made of material that in other circumstances might have become a smaller sister to the Sun. Exciting new data have been gathered by the recent Jupiter probe. Carrying sensing instrumentation and transmitting equipment, the probe descended by parachute through the giant planet's outer atmosphere. As it descended, it sampled its chemistry before increasing heat and pressure finally terminated its signal. Two contending views had dominated discussions about Jupiter for many years: one group of astronomers assumed that Jupiter had a Sunlike composition, while a much larger group contended that Jupiter had been enriched by water, oxygen, carbon, and other elements through a more planetlike accretion of materials. Surprisingly, the Jupiter probe did not find the concentrations of water most astronomers had expected. Instead, the new data suggested that Jupiter's composition was more like that of the Sun. (Stay tuned, we've still got lots to learn.)

## OTHER STARS, OTHER PLANETS

About one hundred years ago, astronomers speculated that the solar system was created after a close encounter between the Sun and another passing star, which pulled material out of the Sun to form the planets. Sad to say, such a scenario would make planetary systems extremely rare in our galaxy. The distances between most stars are huge, and the probability of close encounters is minuscule. Another notion, first suggested by Immanuel Kant in 1755 and refined by Pierre-Simon de Laplace, proposed that planets form naturally from the condensation of a nebulous dust cloud circling its star. If we include the origin of the Sun itself in the condensation process, a broad narrow disk of material should form as a consequence of its initial spin, followed by gravitational contraction. This hypothesis conforms nicely with the fact that all the planets revolve around the Sun in the same direction as the Sun spins, and in a very narrow plane. Such a star-formation scenario is also consistent with the prevalence of double- and triple-star systems. It also implies that all solitary stars may be accompanied by a family of planetary satellites.[13]

Claims for discovery of planetary systems circling nearby Sunlike stars in the last few years have been met with much acclaim, even appearing on the cover of *Time* magazine.[14] Actually, none of these giant planets has been seen directly; they have been inferred from careful spectral analysis of their parent star. Too small, too distant, and too dark to be seen against the bright glare of their star's image, nearly all of these giant planets have been detected by slight "wobbles" in the regularity of movement of their parent star. The wobble is due to both the star and its planet rotating around the center of their common center of gravity. The more massive the planet, the greater its effect on the system's center of gravity and the greater the star's apparent wobble. In the case of our solar system, Jupiter causes the Sun's speed to vary by about twelve meters per second over a twelve-year period. These slight changes in movement can be sensed by minute shifts in the spectra of the star's light. Just as the whistle of an approaching train changes its pitch as it approaches, passes, and moves away from us (known as the Doppler effect), light frequencies are shifted by the movement of objects that emit them. (One of the very strange and invariable constants in our universe is the speed of light. The light coming out of your flashlight is *always* traveling at the same speed; it makes no difference if you're standing still or traveling thousands of miles per hour. But the wave frequencies do change with the speed of your flashlight and whether it's coming or going; this again is the Doppler effect.)

Current instrumentation allows us to sense changes in stellar velocity as low as about three meters per second. If the planets of a distant star are orbiting in the plane of our sight, the star will appear to be moving to and fro as detected by the Doppler effect on their light spectra. However, if our line of sight is at right angles to the orbital planes, the star will appear to move slightly from side to side—not easy to detect from so far away. And if we are viewing that distant orbital plane from an angle, there will be a combination of both movements. Detecting recurring minuscule irregularities in the spectra and motions of distant stars is an amazing feat of new instrumentation, precise long-term observations, and sophisticated statistical analysis.[15]

Though difficult to detect, the search for wobbles in the movement of nearby Sunlike stars resulted in a major discovery in 1995 when the star enumerated as 51 Pegasi gave evidence of having a planetary companion. However, this discovery presented astronomers with a real surprise. From its gravitational effects 51 Pegasi's companion planet is estimated to be half the mass of Jupiter, yet circling its star every four days! Such a quick rate means that it is circling its parent star at a distance much closer than Mercury's orbit around our Sun. This heavy planet wasn't anywhere near where one would have expected to find a big planet. A little later, a second planetary analysis indicated a planet eight times heavier than Jupiter orbiting its parent star, 70 Virginis, within a distance equivalent to the orbit of Venus. The stars chosen for such intense analysis are similar to our Sun in size and temperature. As of early 2001, giant planets have been inferred for about fifty Sunlike stars. Thus, the notion that planetary systems might be common around Sunlike stars has been strongly affirmed.

Among these recent discoveries is a giant planet circling a star very similar to our Sun, 16 Cygni B. This particular planet, 1.6 times as heavy as Jupiter, is unusual in having a very eccentric orbit, ranging in distance from its star from 0.6 to 2.7 astronomical units. (An astronomical unit or AU is the average Earth-Sun distance, about 93 million miles or 150 million kilometers.) Taken together, these newly discovered planetary configurations have forced astronomers to make profound revisions in their theories of solar-system evolution. "Hot Jupiters," close to their stars, have been especially difficult to understand, as have been the highly eccentric orbits of a number of these giant planets. Powerful gravitational effects of a young coalescing star should have incorporated any close-by Jupiter-like planet within the star itself. To get around this dilemma, theorists have developed a variety of hypotheses. Some of these scenarios suggest that such Jupiter-sized planets were formed more distant from their star but were forced into closer or eccentric orbits because of complex interactions with sibling giant planets or by a swarm of smaller planetesimals in a kind of combat for stable orbits. (Larger Jupiter-sized planets, it had been thought, are most likely to be formed about five astronomical units distant from their condensing star, as is Jupiter. At such a distance they can build up their bulk by accretion of material that is too far from the parent star to fall into the star itself.)

The newly discovered extrasolar planets seem to fall into two groups. A number of them circle very close to their stars; these are the hot Jupiters. The others are more distant from their star, but many have eccentric orbits. Either way, this is bad news for those hoping to make new friends on planets circling nearby stars. The hot Jupiters, as they drifted inward, would have swallowed up all nearby planetary material or sent it flying out into space. Highly eccentric orbits of the more distant massive planets should have had similar effects. Both scenarios eliminate the possibility of smaller rocky planets at a distance likely to support active life-forms. All these hypotheses imply a further conjecture: a solar system with stable inner planets may be possible only in those systems with a more distant *and* unperturbed large Jupiter-like planet with a near-circular orbit. And wouldn't you know, that just happens to be the situation down here in our own dear solar system.

Finally, in 1999 astronomers reported "directly visual" data for two of the extrasolar planets. One report detected a slight diminution in light as the large planet passed in front of its star, apparently directly in our line of sight. The other claims to have detected glimpses of increased light, reflected by the massive planet as it swung around its star every few days. Unfortunately, our methods of analyzing the wobble in distant stars can detect only very massive planets that are reasonably close to their star. Planets similar to the Earth are just too small to produce detectable effects on their star, and so the search for planets orbiting nearby Sunlike stars continues to be restricted to heavy planets. In addition, astronomers seek to document at least two good wobbles to feel confident that they have indeed found a massive planet. Thus, to identify a planet like Jupiter, which takes twelve Earth years to swing around the Sun, requires twenty-four years of good data.

In summary, recent discoveries of massive planets orbiting nearby stars have been significant for three reasons. First, they have given us strong confirmation that planetary systems are common around solitary Sunlike stars. Second, none of these systems are identical, exhibiting a variety of configurations that no one had expected or imagined. Third, they suggest that planetary systems with a Jupiter-mass planet in a dense field of planetesimals (or other source of perturbation) may be subject to complex interactions, causing the massive planet to become eccentric or to move inward toward its star. This would sweep up or eject smaller Earthlike planets in its path. Luckily for us, our premier planet, Jupiter, remained in a distant and stable, almost perfectly circular, orbit over the history of our solar system. One conclusion is inescapable: stars with hot or eccentric "Jupiters" are utterly unlikely to be the home of distant civilizations. Surveying the year's scientific advances in the 17 December 1999 issue of *Science*, the editors conclude: "Although the string of new data confirms that planets are common, at this stage it seems that planetary systems configured like our own are rare indeed."

## A VERY SPECIAL SOLAR SYSTEM

Just as most of the newly inferred planetary systems differ significantly from each other, our solar system may be a very unusual configuration. Mercury, Venus, Earth, Mars, and members of the asteroid belt are made up of high-density rocky materials. Earlier, theoreticians suggested that the accumulation of solid material and formation of rocky inner planets should be commonplace in planetary systems because more volatile material will have been blown off or burned away from all those bodies close to the star during early stages of star formation. Of all the planets, Mercury has the densest mass; perhaps its outer, lighter crust was blasted away early in the history of the solar system, when collisions with large asteroids or comets were common. Many large craters on Mercury, Mars, and the Moon are evidence for such a period of heavy bombardment. And let's not forget that rocky planets of substantial size are only likely to form if the condensing presolar nebula is one that is rich in the heavier elements. Significantly, nearly all the recently detected stars whose trajectories appear to be wobbling, thanks to the presence of a massive planet, are stars with metallicity as high or higher than that of the Sun. This brings up the strong possibility that these stars with higher metallicity got that way after swallowing their rocky inner Earthlike planets—planets thrown inward as their hot Jupiter drew closer.

Beyond the rocky inner planets and the more distant asteroid belt, the composition of our solar system changes dramatically. Here four giant planets sweep majestically around our star. The first two are the largest: Jupiter and Saturn. They are thought to have begun with dense cores, perhaps ten times as massive as the Earth. These cores aggregated a huge volume of gas early in the history of the solar system, and that gas makes up a huge percentage of their volume. The more distant duo, Uranus and Neptune, are smaller, having formed in a region where the solar nebula was less dense. Because of high percentages of hydrogen and helium, our four large outer planets do not have the rocklike density of the small inner planets.

Beyond the large outer planets of our solar system lies another ring of objects, small and much more difficult to detect. Varying from the size of comets (1–20 km and common) to the size of small planets (2,000 km but rare), these objects range beyond the orbit of Neptune out to about 50 astronomical units from the Sun. They and the short-period comets (cycling around the Sun in less than 200 years) are part of what is called the Kuiper Belt. All the members of the Kuiper Belt orbit the Sun in the same narrow plane in which the planets and asteroids travel. In contrast, the more distant long-period comets (cycling through the solar system in orbits of over a thousand years) have many different trajectories and are not so confined. They seem to come from a spherical region called the Oort Cloud, 1,000 to 20,000 astronomical units away from the Sun. Comet Hale-Bopp, so spectacular in early 1997, is a member of the Oort Cloud; it should return in the year 4380.

Comets, whether from the Oort Cloud or the Kuiper Belt, are often referred to as "dirty snowballs." Their surface reflectance suggests that they contain simple compounds, such as water, methane, ammonia, carbon monoxide, and hydrogen sulfide, all in frozen form. Because temperatures cool enough for hydrogen and oxygen to form water-ice existed only beyond the orbit of Jupiter in the early solar nebula, it seems reasonable that much of the water in today's solar system is held within the distant comets. Once formed, however, many comets were tossed out of the region of their origin into the distant Oort Cloud by the gravitational effects of the giant planets. This also scattered them within the solar system to clobber the inner planets during the early eons of our history.

Distant little Pluto is the smallest planet, having only $\frac{1}{2000}$ the mass of the Earth. Pluto and Triton (the larger of Neptune's two moons) were probably icy members of the Kuiper Belt, having since been captured by the gravitational forces of the large outer planets. In contrast, a few of the denser rocky moons of Jupiter and Saturn are likely to have been members of the inner asteroid belt. Located between the orbits of Mars and Jupiter, the asteroid belt contains thousands of smaller rocky objects that are the source of tiny meteorites that often illuminate our night sky as "shooting stars."

Except for the nearest and the farthest (Mercury and Pluto), the planets orbit the Sun within 4 degrees of the Earth's orbital plane (the ecliptic), and that plane differs by only 7 degrees from the plane of the Sun's rotation. Planetary orbits are spaced farther apart the more distant they are from the Sun in a loosely approximate mathematical sequence called the Titius-Bode rule (take the series 0–3–6–12–24 . . . , add 4 to each, and divide by 10 to get the approximate distances of the planets in astronomical units). Interestingly, this mathematical series had a calculated orbit between Mars and Jupiter that *lacked* a resident planet. This puzzle was soon unraveled. Through careful searching, the asteroid belt was discovered at the orbital distance calculated for the "missing planet."[16] Asteroids circle the Sun in the same narrow plane as do the inner and outer planets. Shortly after their discovery, the asteroids were thought to be the debris of a planet that had disintegrated, but meteoritic studies suggest that the asteroids are ancient material that failed to coalesce into a single larger planet. The two small moons of Mars, only about six miles (ten km) in diameter, are very likely captured asteroids.

Despite the diversity in size and composition of all these many planets and moons in our solar system, it is humbling to be reminded that the entire array of objects orbiting the Sun, including us and giant Jupiter, amounts to only 1 percent of the mass of the solar system. On a more positive note, the planetary and cometary entourage circling our star *does account* for 97 percent of the angular momentum in the solar system.

If we interpret some of the moons of the solar systems's giant gaseous planets as the products of gravitational capture from either the Asteroid or Kuiper Belts, the original solar system appears to have had a more simply structured configuration. The innermost planets and asteroids are made up of rocky materials with a great variety of both lighter and heavier elements. The four large, distant planets arrange themselves into two groups: gaseous Jupiter and Saturn, and slightly more icy Uranus and Neptune. The icy comets formed near the orbit of Jupiter and beyond. In this simplified view of the solar system, little Pluto, with a very eccentric orbit, is nothing more than an interloper from the Kuiper Belt and quite different from the Sun's other planets, while the outer comets have been strewn about by the gravitational effects of the giant planets. These regularities in size and composition in our solar system stand in contrast to the diversity displayed by the sixty satellites that circle the planets themselves. A violent early history with many collisions and complex interactions seems to be the only explanation for so diverse a zoo of satellites, no two of which are identical.

## THE GOLDILOCKS ORBIT

Life is a dynamic process that can exist only where energy arrives at a steady rate in a form that is useful and with an intensity that is neither too strong nor too weak. Such a situation is likely to be found only near a stable, brightly burning star. For the possibility of life, the presence of rocky inner planets, not too close to the star and not too distant, is crucial. One of these inner planets must be located in what astronomers have dubbed the "Goldilocks orbit." Heroine of a children's fable, Goldilocks was a very fussy little lady. On her visit to the home of three bears (Mama, Papa, and Baby Bear) who happen to be elsewhere, Goldilocks tries the chairs: one is too big, one is too small, and one is just right. She then tries out the bowls of porridge: one is too hot, one is too cold, and one is just right. After filling herself with porridge, she goes up to the bedroom where she finds one bed is too soft, one is too hard, and one is just right, and she proceeds to fall asleep. (The story gets more interesting when the bears return.) For life to prosper, the Goldilocks principle must be satisfied: conditions have got to be *just right*. To sustain complex life-forms anywhere in the universe, it seems likely that a planet has to be at a distance from its parent star that provides a temperature range in which water is maintained in its liquid state. The astronomer James Kasting has calculated that in our solar system, the orbital distances in which water will remain liquid on a planet's surface range from a minimum of 88 million miles to a maximum of 127 million miles.[17] At an average distance of 93 million miles from the Sun, planet Earth sits comfortably within this Goldilocks zone.

However, there are other possibilities for life in our solar system. Europa, Jupiter's moon, may have liquid water beneath a surface of thick ice, kept warm by internal tidal forces generated by the giant planet. But without warm starlight to power photosynthesis, it's unlikely Europa can support anything more than the simplest life-forms. In their book *Rare Earth*, Peter Ward and Donald Brownlee discuss the habitable zone—the region of the Goldilocks orbit—from a number of points of view.[18] They point out that the width of the habitable zone depends on the kind of life one is looking for: versatile microbes have a much broader zone than do terrestrial animals. More significantly, if Sunlike stars grow warmer as they age, as astronomers theorize, the habitable zone must shift outward. Since planets cannot shift outward to remain within the zone, life that originate early in the history of a solar system may be extinguished. Luckily, we escaped that fate.

Other aspects of our solar system may have been critical to the evolution of life on Earth, especially the giant planets. Jupiter's large size and near-circular orbit at a safe distance from the inner planets was an especially lucky coincidence. The inability of a planet to form within the asteroid belt close to Jupiter is likely a product of the gravitational effects of the giant planet itself. Likewise, because Mars is only about half the size of Earth or Venus, some astronomers have suggested that Jupiter's gravitational effects prevented our reddish neighbor from achieving a more substantial mass. Clearly, a Jupiter-sized planet circling our star at the distance of Mars or with a strongly eccentric orbit would have made an Earth-sized planet in our orbit extremely unlikely.

Our comfortably distant giant planets may have had an additional beneficial effect for planet Earth. Computer simulations suggest that gravitational effects by our distant neighbors, the giant Jovian planets, have either swallowed up many comets and asteroids or sent them sailing out of the solar system, reducing the probability of having such space debris smash into the Earth over the last few thousand million years. Think of the spectacular demise of comet Shoemaker-Levy as it broke into fragments and later cascaded into Jupiter with its clearly visible effects in the summer of 1994. The dark splotch left in Jupiter's outer atmosphere by the impact of one of those fragments was larger than planet Earth! Early in the history of our solar system the giant planets apparently did send comets flying every which way, and those comets that ran into our planet appear to have enriched it profoundly (more about that in the next chapter). These collisions are calculated to have been 99 percent behind us by about 3,500 million years ago. Since then, computer models suggest that the "cleansing effect" of the large Jovian planets may have reduced the probability of high-energy Earth-comet impacts from once every 100,000 years to once every 100 million years. One such high-energy impact 65 million years ago may have ended the reign of dinosaurs. One doesn't need much of an imagination to conclude that a planet clobbered by such devastating impacts every hundred thousand years (or so) might be a planet without rain forests or advanced civilizations. Thanks Jupiter and Saturn! Without you we probably wouldn't be here.[19]

After reviewing all these many aspects of our star and solar system, it seems clear that we have an exceptionally felicitous astronomical environment. The Sun's location, size, singularity, metallicity, and planetary configuration have all contributed to the fact that on the blue and white planet, the one in the Goldilocks orbit, one species now probes the deep mysteries of space. And just as our Sun is a very special star, evidence that's both old and new will make clear that this rocky sphere we call Earth is no ordinary planet.

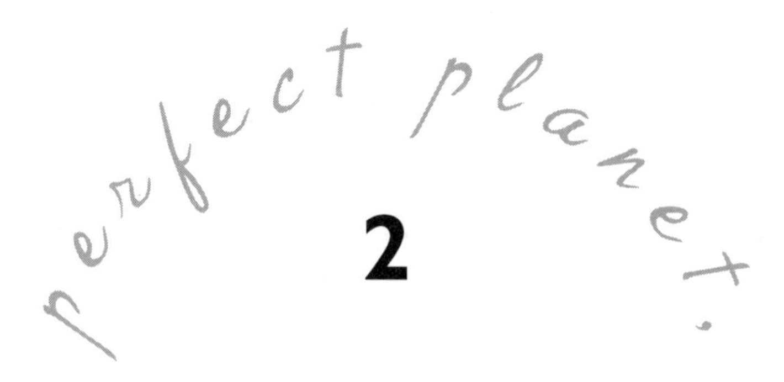


# A REALLY GOOD PLANET IS HARD TO FIND

clever species

The "comfort zone" of temperatures that can support life is a very narrow one, representing but a thin sliver in the middle of our galaxy's huge temperature range. With interstellar space only a few degrees above absolute zero and temperatures of millions of degrees within the stars, environments that can support life are extremely rare in the universe. Complex information-carrying molecules begin to come apart not far above the boiling point of water. Since temperature is a measure of molecular motion, larger molecules, and eventually even simple molecules, will be torn asunder as temperatures increase.

At the other end of the thermometer, as temperatures plunge below the freezing point of water, chemical reactions are slowed and can no longer sustain active complex life-forms. This is true for life as we know it, and these parameters are probably universal. It's likely that all life, of whatever configuration and wherever in the universe, can operate only in a very narrow band of temperatures close to those found right here on planet Earth.

Even more fundamentally, the size of living things seems to be very tightly constrained. Quarks, the smallest significant bits of matter, come in only a few forms, which, when united into protons and neutrons, fashion ninety stable elements. Most of these elements, in turn, can find comfortable energy states by sharing electrons with each other, creating a fantastic diversity of chemical compounds. It is at this size scale where the organization, catalysis, and self-reproductive activities of life become possible. At larger metric scales, one cannot find the diversity of a molecular world from which to construct higher levels of organization. In addition, the larger distances themselves would inhibit quick reactions or the possibility of timely communication. Regardless of the galaxy, life processes are likely to be based on that lilliputian world where molecules quickly interact and transform themselves, and where they are common enough to continuously engage each other. When considering the origins of life, the best recipe would seem to be a warm soup of diverse, closely interacting molecules.

## WATER, THE MEDIUM FOR LIFE

Virtually all of life's basic processes on planet Earth take place within the medium of water, and it is difficult to imagine any form of life beginning in anything except a liquid medium. No matter where you are in the universe, water is probably the most likely liquid medium for life activities. Ammonia and ethyl alcohol may be distant alternatives, but ammonia has a cold and narrow temperature range in which it is liquid: –77 to –33°C, and neither alcohol nor ammonia possesses the wide array of attributes that make water so useful a solvent.

Though not uncommon among the chemical compounds in our galaxy, water in its liquid form is very hard to find; that's because liquid water is stable only under conditions of considerable pressure. In thin interplanetary or interstellar space, water quickly transforms itself directly from crystals of ice (when cold) into gaseous water vapor (when hot), and vice versa. Water is also distinctive in many other ways, with very unusual properties for a compound of its molecular weight. Water is a liquid over a wide range of "comfortable" temperatures and pressures at which other compounds of similar or even heavier weight are gases. It acts as a solvent for a wide variety of both organic and inorganic compounds, allows for the flow of electricity, and yet is relatively inert. Because water molecules are somewhat dipolar (with a positive side and a negative side) they tend to "stick together." This gives water a high density, high surface tension, and a high boiling point, and explains why it remains liquid over a relatively wide temperature range. Its surface tension, together with relatively low viscosity, is essential in moving water through minute pores in soil and to the tops of tall trees. The dipolar nature of water molecules and their tendency to stick together also explain why water absorbs so much energy as it becomes warmer or before changing from liquid to gas. This gives large bodies of water the ability to moderate temperature changes. We use this high "heat of vaporization" ourselves, dissipating excess heat as we perspire.

A number of fundamental physical characteristics give the water molecule its unusual properties. Because the two hydrogen atoms are oriented to one side of the oxygen atom, the water molecule has a positively charged (hydrogen) side and a negatively charged (oxygen) side. In addition, since oxygen is a highly electronegative element, the oxygen atom binds the two hydrogen atoms' electrons more closely to itself within the water molecule. This has the effect of making the projecting hydrogen atoms more positive and enhances the formation of weak "hydrogen bonds" with the negative sides of other nearby water molecules (since positive and negative sides are attracted to each other). These are the structural reasons for some of the unusual properties of the water molecule mentioned above. And, as if that weren't enough, it just so happens that the two hydrogen atoms form an angle of 104.5 degrees with the center of the molecule—very close to the angles of a four-pointed tetrahedron. This little detail means that freezing water molecules build a crystalline lattice with *four* other water molecules, instead of a closer union with six or more water molecules. The tetrahedral crystalline structure creates more open space between frozen water molecules. And this added space explains why ice weighs less than cold water *and floats*. Virtually unique in the chemical world, this property of water has profound ecological consequences. If lakes and oceans were to freeze from the bottom up, aquatic life in colder zones would be severely reduced. Unfortunately, this same property can tear cells apart as water within them freezes and expands; living things require elaborate biochemical protection for survival at low temperatures.

But back to the positive side. Both ice and snow have low thermal conductivity, helping protect small Arctic animals from the subzero winds above and insulating polar waters beneath surface ice. Things might have been even better if ice and snow were dark in color, absorbing light energy instead of reflecting so much of it back into space. The highly reflective quality of ice and snow makes a runaway glacial climate a scary possibility, as warm sunlight is reflected back into outer space.

All fundamental cellular processes occur within the liquid medium of water. This is true whether capturing the energy of sunlight, "burning" hydrocarbons to provide energy within the cell, building and repairing DNA, or facilitating movement of chromosomes before cell division. Functioning as the basic medium of living processes, cell interiors average between 70 percent and 90 percent water. The ability of water to break up into positive ($H_2OH+$) or negative ($OH–$) ions allows it to dissolve a great variety of substances and promotes many chemical reactions. The tendency of water molecules to form an oriented layer or "cage" around nonpolar substances helps shape and stabilize larger molecules such as proteins and nucleic acids. For these reasons, life as we know it can occur only within temperatures and pressures at which water remains in its liquid state.[1]

## CARBON, THE ESSENTIAL STRUCTURAL ELEMENT

Science fiction authors once speculated about the possibility of life based on silicon, a sister element to carbon. But one doesn't hear much along those lines nowadays. At temperatures where water is liquid, silicon unites with oxygen to form insoluble quartz, the most common mineral in sand and granite. In contrast, carbon combines with oxygen to form either carbon monoxide or carbon dioxide, gases that are soluble in water, where they produce a weakly acidic effect and the buffering effects of carbonates. Also, silicon is not as reactive or as adept at forming chemical bonds as is carbon.

Flanked in the periodic table by elements that are more energetic in grabbing or donating electrons, carbon has a unique versatility, able to share its four outer electrons with many other elements, as well as with others of its own kind. The smallest element that can share the maximum number of four electrons, the carbon atom is able to form strong covalent bonds with as many as four partners. No other chemical element is so adept at forming chains, rings, and other convoluted molecular shapes. Thanks to the ease with which it can form strong bonds with hydrogen, oxygen, nitrogen, sulfur, and itself, carbon is the central building block in those molecules that make up living systems here on Earth. Between 10 and 20 percent of the weight of living things is made up of carbon. Considering its unusual flexibility in forming chemical bonds, carbon is the only reasonable candidate for life's central building material throughout the universe.

Carbon dioxide, like water, has many unusual properties with important positive effects for life on Earth. It is unusual among gases in maintaining a similar number of molecules in water solution as in the air at ordinary temperatures—a critical feature for photosynthesis on land and in the sea. Also, its ability to quickly reverse from aerial gas to dissolved gas and to form carbonic acid in water helps stabilize water acidity. Being transparent to ultraviolet light but absorbing the weaker infrared wavelengths makes carbon dioxide a critical part of the aerial blanket insulating our planet from the loss of heat to the near-vacuum frigidity of outer space. High-energy ultraviolet easily penetrates the atmosphere; it is absorbed by surface materials and partly reradiated as lower energy infrared radiation. It is this reradiated heat energy that carbon dioxide, water vapor, and other "greenhouse gases" absorb, helping to maintain a warm, habitable surface on our planet.

Consistent with their importance in the chemistry of life on Earth, hydrogen, carbon, nitrogen, and oxygen are among the most common elements in our galaxy. Carbon is the fourth most common element in the universe. (However, bear in mind that, together, hydrogen and helium add up to more than 98 percent of the atoms in the galaxy.) But that's enough chemistry; let's move on to planet Earth.

## THE RIGHT DISTANCE, THE RIGHT SIZE, THE RIGHT SPIN

Life is dynamic and demanding, requiring a constant stream of energy for its sustenance. Only where powerful radiation streams outward from an intense but stable source into the void of dark and frigid space can we expect to find that precise energy flux that will power living things. This form of energy must not be too intense, or it could destroy the delicate fabric of living tissue. Nor can it be too weak, incapable of energizing the reactions that power active living beings. Such special requirements are likely to be found only near a brightly burning Sunlike star; and it is only at a critical distance from that star where these conditions will be "just right."

Our planet is very special in the solar system. It sits at just the right distance from the Sun to maintain water in liquid form over most of its surface. As we saw in the first chapter, this is what astronomers are calling the Goldilocks orbit. One calculation has suggested that the Goldilocks orbit in our solar system ranges from 95 to 102 percent of the average Earth-Sun distance, but such a narrow range seems questionable. A lot depends on the nature of the atmosphere: too thick and a hothouse may develop; too thin and a permanent frigid glaciation may result. Earth has both the right orbital distance *and* the right atmospheric qualities.[2] Just as important, Earth's orbit is close to being circular. Our distance from the Sun differs by only about 3 percent during the year, resulting in a maximum energy difference of about 7 percent. Fortunately for the Northern Hemisphere, we are currently closest to the Sun in January and most distant in July; otherwise northern winters might be even colder.[3]

An additional factor of critical importance for the development of life on Earth has been our planet's quickly paced spin. It's fast enough so that days and nights do not reach extreme temperatures. If days and nights were measured in weeks instead of hours, we would have temperature swings of hundreds of degrees and life on land would be impossible. Think of Mercury where the "day" lasts for about two-thirds of its yearly voyage around the Sun, resulting in a torrid side and a frigid side. Venus also rotates very slowly, but a dense atmosphere whirling around the planet keeps the entire surface frying under a toasty blanket. At the other extreme, if the Earth's day was only a few hours long, gale-force winds (produced by the rapid spin) would likely make the evolution of complex terrestrial vegetation impossible.

Another nifty feature of our planet is its axis of spin, tilted about 23 degrees away from being perpendicular to the plane of our orbit around the Sun. This tilt brings the Northern Hemisphere longer days in May to August and shorter days in November to February, producing our annual cycle of seasons. Without this annual cycle, the boreal forests and tundra of the North would be uniformly frigid, supporting far fewer plants and animals. Even more important, the tilt of our axis shifts weather systems back and forth, north and south, across tropical latitudes in a consistent annual cycle. These annual weather patterns, always in line with the Sun, sweep monsoonal rains back and forth across the seasonally dry tropics, covering a much wider swath than if our spin-axis was consistently perpendicular to the orbital plane. A slight tilt in our axis has contributed significantly to the richness of the world's terrestrial biota.

Our planet's mass is another significant factor in making planet Earth hospitable to active terrestrial life forms. Could sprightly land animals have evolved in a gravitational field substantially stronger than Earth's? Probably not. A much stronger gravitational field would preclude the muscular agility and fast responses associated with the quick wits of really smart land animals. At the other end of the scale, a weaker gravitational field cannot prevent the slow evaporative loss of important gases, as has occurred on Mars. All higher animals require considerable oxygen pressure to carry on active respiration, and this requires an atmosphere kept sufficiently dense by a gravitational field similar to that of planet Earth. In addition, some astronomers suggest that smaller planets might not be able to remain warm in early stages when a Sunlike star would be cooler. Others think that a larger planet might form too dense an atmosphere and get trapped in a torrid greenhouse effect.[4] Though we should remain skeptical of such grand speculations, it seems highly likely that intelligent life-forms capable of building radio telescopes can evolve only on planets not much different in mass and spin from the planet on which we happen to be sitting.

## OUR MOON, MORE THAN JUST A PRETTY FACE

Except for dark and distant little Pluto, Earth is unusual in having the largest moon, proportional to its size, in the solar system. Fifth largest in the solar system, our Moon is a bit more than a quarter of the diameter of the Earth. Our big satellite produces a terrific stabilizing force, minimizing the long-term wobble of our axis and helping maintain the regularity of our seasons over hundreds of thousands of years. Without such stabilization, vegetation zones might be subject to far more severe long-term change. However, our axis does wobble, from 21.8 to 24.4 degrees over a period of 41,000 years. A recent analysis of Mars has suggested that the red planet's axis may have shifted its tilt from 15 to more than 40 degrees over periods as short as ten million years. The fact that the present tilt of Mars is very similar to that of Earth is, according to this information, simply a matter of accidental timing as Mars changes its tilt over millions of years. More dramatically, one calculation suggests that Earth's axis, without our moon, might swing between tilts of 0 and 85 degrees. A slow but major wobble of such magnitude in our spin-axis might not be too troublesome for life within the ocean, but for life on land it might be disastrous. If the obliquity of Earth's spin-axis had shifted dramatically over millions of years, changes in day length, seasonality, and rainfall patterns would have created great stresses for life on land. Without its large stabilizing Moon, the Earth would have suffered waves of extinctions over many land areas. More than a beautiful object of the night, our large satellite has provided Earth with a strong steadying influence as we spin about our axis, wobbling just a little.

Because the Earth "bulges" ever so slightly in response to the Moon's gravitational pull as we spin around each day, our planet is subject to a kind of tidal friction. This tidal drag has played a role in slowing down Earth's rate of spin. Days may have been as short as ten hours when the Earth was young, 4,000 million years ago. By 900 million years ago the length of the day is estimated to have been about eighteen hours. Thanks to the tidal friction of both the Sun and our big satellite, continued slowing has brought our daily spin down to today's twenty-four hours. Without the Moon, a faster-spinning Earth would have had two negative effects, argues Neil Comins in his book *What If the Moon Didn't Exist?*[5] First, climatic patterns would be more tightly constrained by restricting rainfall patterns to narrower bands around the planet. Second, higher winds would have made the evolution of tall land plants far more difficult, perhaps eliminating the possibility of ever developing a lineage such as agile and dexterous tree-climbing monkeys.

Early in our planet's history, the Moon may have benefited the Earth in an even more fundamental way. Many theories have been put forward to explain the unusual Earth-Moon pairing. But the only scenario that seems dynamically plausible and consistent with the differences between the Earth's and the Moon's surface chemistry is a violent early collision: the "big whack" theory. Other, less violent, theories of Moon capture are much less successful in explaining the Earth-Moon relationship. The big whack hypothesis claims that a grazing impact with a Mars-sized body injected heavy material into the Earth, while vaporizing much of the incoming object. The vaporized material, catapulted into Earth orbit, would later condense anew as our satellite. Such a scenario is consistent with both the lack of a larger metal core at the center of the Moon and the fact that the Moon's orbit around the Earth is nearly circular. (Capturing a satellite by close-encounter should have produced an elliptical orbit for the Moon.) Latest estimates place the big whack at between 4,520 and 4,500 million years ago, perhaps as the Earth was still becoming a consolidated planet.[6]

There are several important aspects to an early Earth–proto-Moon collision. The violent impact may have contributed heavy metals to the Earth's core and made a significant contribution to our planet's mass. This should have had the effect of giving the spinning Earth a stronger magnetic field (the magnetosphere). This is a force field helping deflect the solar wind, which might otherwise strip away the stratospheric ozone layer shielding us from high-energy ultraviolet radiation. This was particularly important for the later evolution of life on land. Second, there is the possibility that the big whack helped give our planet its zesty spin. Think about the nearly yearlong spin of our sister planet, Venus. If the Earth's day was measured in weeks or months, complex life forms could never have populated land surfaces. Third, the whack may have created the tilt which gives us the seasons that enhance biological diversity in many parts of the globe. And finally, the collision may have had a much more immediate and important effect long ago: ridding the early Earth of a suffocating atmosphere. By blasting a dense blanket of carbon dioxide and methane into space, the Moon may have saved planet Earth from Venus's deadly fate.

## VENUS, A VERY DIFFERENT PLANET

Venus had always been a beautiful mystery. Often the brightest "star" in our firmament, it has been the subject of myth and speculation ever since humans first contemplated the night sky. Venus is our sister planet, almost identical in size and density. Covered by thick clouds, it was once thought of as a possible home for luxuriant tropical life. But the first accurate radio telescope estimates of surface temperatures on Venus were disheartening: they indicated temperatures far above the boiling point of water. Our government supported an ambitious program in the 1970s to study Venus with satellites and space probes. One of the arguments used by planetary scientists to justify spending hundreds of millions of dollars on Venus missions was that learning about our sister planet would help us understand our own planet. They knew Venus was too hot to harbor life, but they thought that new data from our sister planet could tell us something about Earth's early geological history. A rich harvest of new data, especially the imagery produced by the Magellan orbiter radar scanner, has made clear that these early aspirations were very wide of the mark. Beyond size and weight, Venus bears little resemblance to Earth.

Both spectral analyses and Russian space missions have shown that Venus's dense atmosphere helps maintain an ovenlike surface temperature of about 900°F (475°C)—the result of a runaway greenhouse effect. This effect begins with high-energy radiation from the sun penetrating the atmosphere and heating the planet's surface, then reradiating outward as lower-energy infrared radiation. Though the high-energy wavelengths easily penetrate Venus's thick atmosphere, the reradiated lower-energy infrared wavelengths, trapped by dense clouds of carbon dioxide, cannot escape back into space. An orbit closer to the Sun, together with a thick suffocating atmosphere, has turned Venus into a sterile furnace. As if that weren't bad enough, daylength on Venus is slightly longer (243 Earth days) than the time it takes the planet to revolve around the Sun (225 Earth days). With one side facing the Sun for many months, and the opposite side facing the blackness of night, Venus would have a really hot side and a really frigid side, were it not for its dense atmosphere. Even without a suffocating environment, this is a planet that could not support living things.

Sophisticated cloud-piercing radar from the orbiting Magellan spacecraft has given planetary scientists a detailed picture of the surface topography on Venus. Not only is that planet inhospitable to any form of life, it is geologically moribund. An important finding has been the essentially random distribution of impact craters on Venus. This means that no areas have been covered by fresh new lava or ash falls for hundreds of millions of years. It is easy to see that the cratering on our Moon is not at all random. Broad darker areas with few impact craters on the Moon (called "maria") are the result of extensive lava flows, contrasting with the grayish and older crater-covered areas. The sparse but uniform density of cratering on Venus suggests a surface between 500 and 900 million years old. It seems that this planet, in what may have been a colossal convulsion, developed a new surface at that time, and the outer crust has been uneventfully dead ever since. Our sister planet, once thought to harbor tropical forests, has turned out to have a high-pressure atmosphere of poisonous gases baking a rigid, torrid surface. No longer the subject of science fiction adventures, Venus is now known to be one of the most inhospitable members of the solar system.

An unexpected bonus of the exploration of Venus and other planets has been recognition of one of Earths's most unusual features: a continuously active crust. No other planet or moon in the solar system has a dynamic system of mobile plates, and, more important for our story, none has a surface of two significantly different elevations.

## EARTH'S CONTINENTS: STABLE OR MOBILE?

The notion that continents had moved across the surface of the Earth was championed by Alfred Wegener in the early decades of the twentieth century. Wegener used many lines of evidence to support his theory of "continental drift." The obvious symmetry of the eastern and western shores of the southern Atlantic ocean, with matching geological strata in a few areas, was an important part of his argument. Even a child can see the general symmetry of the Atlantic's eastern and western shorelines on a globe. But the obvious symmetry was meaningless for those geologists who adhered to a theoretical framework that considered continental movements impossible. As a climatologist and unbeholden to the "stabilist" thinking that characterized the geological worldview, Wegener mapped out a bold conjecture.

One of Wegener's arguments for continental movement was based on the fossil *Glossopteris* flora, and a hypothetical ancient Gondwanaland on which these ancient forests grew. An imaginative science fiction writer would be hard-pressed to think up a better name than Gondwanaland, based on Gondwana, a small town in southern India. One of the more important localities where fossil remains of the *Glossopteris* flora are found, Gondwana rocks include formations dated between 280 and 240 million years old. What made the fossil *Glossopteris* flora significant is that it was also found in South America, South Africa, and Australia. If one or two fossil species were the same, one might argue that they were widespread at that ancient time. What contradicts this argument is that *Glossopteris*, a tree with narrow elliptic leaves, is consistently associated with a suite of additional plant forms, each differing from the other and distinctive—found in *each and every one of these* ancient fossil assemblages. Such a strict concordance of distinctive elements in the *Glossopteris* flora strongly implies that the fossils represent samples of what once was a single continuous ancient forest. Based on these nearly identical fossil assemblages, and animal fossils as well, Wegener and a few others concluded that there had to have been a single large southern landmass about 250 million years ago, calling it Gondwana or Gondwanaland. This was a landmass which had, according to the drift hypothesis, been torn asunder to produce the world we know today.

Wegener's general theory of continental drift was the subject of serious discussion in Europe and among scientists living in the southern parts of the world. However, for almost half a century, it was given little credence by geologists in the United States. At a national meeting in 1926, American geologists essentially declared Wegener's theory outside the realm of acceptable scientific discourse. It is easy to understand the negative response by many academics; Wegener's theory is a violation of the general scientific rule of parsimony or "Ockham's razor." In many of his writings, William of Ockham (ca. 1285–1349) argued for minimizing the number of factors used to formulate explanations, an idea that has been a cornerstone for scientific analysis ever since. Mobile continents were an additional and complicating explanatory hypothesis that American geologists did not accept.

It is worthwhile considering Wegener's idea from the perspectives developed by two prominent commentators on the nature of the scientific enterprise. Karl Popper claimed that science tests theories; in his view institutional science may set the basic conventions but scientists theorize on their own. Thomas Kuhn, on the other hand, suggested that science tests the scientist; institutional science validates theories as acceptable at a given time and identifies puzzles worth solving. From Popper's perspective puzzle solving was trivial or poor science, while Kuhn claimed that puzzle solving within the context of currently accepted theory is a regular day-to-day scientific activity. Popper maintained that theories are abandoned when falsified, but Kuhn pointed out that theories were usually retained until better theories came along. For Popper science should welcome unpopular hypotheses, but Kuhn suggested that, in the real world, science treated variant hypotheses as unwelcome.

As a philosopher seeking to distinguish science from other human activities, Popper claimed the testing of falsifiable hypotheses is what made science different. He saw science as creating a progressive evolution of knowledge, where new ideas were subjected to a kind of "attrition by falsification." As a historian of science and more interested in the sociology of science, Kuhn focused his attention on the idea of revolutions in science, when a long-held theoretical framework or ruling paradigm is torn down and replaced by a new one. But why should larger theoretical paradigms require a revolution to bring them down? Others had pointed out that core theories or paradigms quickly became "protected" by subsidiary theories and hypotheses. In part, this is a necessity for helping develop a theory in its early stages. But once a major theory develops a cocoon of subsidiary theories, it becomes very difficult to falsify. Soon the theory, along with its supporting structures, becomes a ruling paradigm that only a "revolution" can bring down. While Popper was correct in arguing that much of science is a slowly evolving process, the theory of plate tectonics provides a superb example of a major revolution in science, just as Kuhn had suggested.

Do continents move? Did they drift apart? Wegener claimed that they did in a well-crafted book, first published in 1915 and revised during the 1920s.[7] But for North American geologists the core assumption could not be challenged: the continents were stable. After all, how could continents plow their way across the solid ocean floor? Finally, the ruling paradigm of continental stability collapsed because new data proved it false.[8]

An early indicator of trouble with the concept of fixed continental positions came with analyses of the position of the north magnetic pole through time. Based on the evidence from ancient magnetism, frozen solid as molten lavas crystallized, it appeared that the north pole had wandered through time. The likelihood of the Earth's spin-axis shifting dramatically over time made no sense, especially considering the enormous mass of the Earth and the stabilizing effect of our large satellite. Worse yet, with new findings, ancient rocks indicated that the north magnetic pole was not in the same place at the same time. Magnetized rocks of the same age, but on different continents, were pointing in different directions.

Meanwhile, careful depth soundings of the Atlantic Ocean had disclosed a huge and continuous undersea volcanic ridge running down the middle of the Atlantic, all the way from Iceland in the north to near Antarctica in the south. Not only was this huge undersea mountain range perfectly parallel with the ocean's eastern and western shores, it was almost exactly halfway between the two shores. This discovery, in itself, had little effect on geological thinking. However, further detailed magnetic analyses, traversing the mid-Atlantic ridge and deep sea ridges in the Indian and Pacific Oceans, soon produced an amazing revelation.

## GEOMAGNETIC REVERSALS

Molten rocks, as they cool and harden, "freeze" not only the direction of the Earth's magnetic field but also its polarity. For reasons that are not understood, the Earth's magnetic field flips over (reverses) from time to time. Magnetic north becomes magnetic south and, after a few to many millions of years, flips back again. These are called geomagnetic reversals and, at first, their discovery was met with disbelief. Requiring careful techniques to document, the reversals have proven very useful in helping date magnetizable rocks that were once molten. The last such reversal was 780,000 years ago. In the mid-1960s, ocean-floor surveys showed that bands of hardened lava paralleling the deep ocean ridges exhibited geomagnetic reversals that were mirror images of each other. In addition, the age of rocks increased in direct proportion with their distance from the ridge, both east and west of the ridge. No one had expected anything like this. Here was the evidence that proved Wegener correct. The magnetic reversal data and accurate dating made it clear that new volcanic magma was arising from great depth and coming to the surface at the *center* of undersea ridges. By moving laterally and forcing older rocks to move eastward and westward, ocean floor spreading at the mid-Atlantic ridge was forcing the shores of the Americas and Europe/Africa ever farther apart. The mid-Atlantic ridge created the Atlantic Ocean basin and is continuing to enlarge it. Our early ancestors had thought the Earth was flat; then we learned it was spherical; now we've discovered that its land surfaces are waltzing around.

With the realization the sea floor of the Atlantic and other oceans had spreading centers, and that some regions of the Earth's crust were being forced underneath advancing margins, geologists developed the theory of plate tectonics to describe a dynamic new view of our planet's surface. An ocean floor nowhere older than 250 million years, marked by midocean ridges and deep sea trenches, established the fact that the ocean basins were the youngest and most dynamic part of the Earth's surface.

One of modern science's major paradigm shifts, plate tectonics was not the product of a well-focused progressive research trajectory. Rather, the new insights were based on newly developed instrumentation to measure faint magnetic signals, the potassium-argon technique for measuring geologic time more accurately, and, finally, shipborne magnetic surveys and cores of deep ocean rocks.[9] Far richer in detail and process than Wegener's proposal, the theory of plate tectonics has given us many new explanatory insights. This fundamental revolution in the earth sciences helps us understand not only how the Atlantic Ocean originated, and why the north magnetic pole seemed to be at different sites when analyzed in ancient rocks, but also how ancient Gondwanaland came to be torn asunder. The new theory proposes that the subcontinent of India, ripped loose from Gondwanaland—with Gondwana on board—rafted northward across the Indian Ocean and slammed into Asia about 55 million years ago. In this way the new paradigm also explains the origin of the world's highest mountain ranges, the Himalayas. The new view also explains volcanic activity and earthquakes along areas of plate collision, such as the "ring of fire" bordering much of the Pacific Ocean. More fundamentally, plate tectonics explains why the Earth, alone in the solar system, has a two-tiered surface.

The Earth's ocean basins average about 12,400 feet (3,800 m) in depth, while the continental surfaces average about 2,200 feet (680 m) above sea level. This 14,600 foot difference is an anomaly in the solar system. Granitic continental rocks are lighter and thicker than the heavier but thinner volcanic (basaltic) ocean floor. Both float on an underlying mantle that is denser than either granite or basalt. Another major difference between oceanic and terrestrial geology is age. While the oldest parts of the ocean floor are about 250 million years old, there are outcrops in southern Greenland, South Africa, and western Australia exceeding 3,000 million years in age. The "youth" of the ocean floor is due to the upwelling of new materials from along the midocean ridges, which crisscross the ocean basins like the seams of a baseball. When a landmass is rammed by expanding ocean floor it tends to ride up and over the heavier oceanic crust, creating mountain chains and maintaining its elevated position. These processes give Earth a dynamic two-tiered surface unlike any other in the solar system.[10]

The significance of plate tectonics is not just pretty mountains. Because of the dynamic crust we have critical resources other planets lack. Active plate margins, volcanic fissures, and oceanic hot spots are the sites where recycling molten rocks can create ores of highly enriched elemental composition. Uncommon elements, heated and recycled at subducting plate margins, have become concentrated ores over millions of years. Eventually elevated to the land surface, such sites have become our most important mineral deposits. The plate tectonic revolution has helped us understand not only the processes that might lead to the concentration of rare elements, but also where to look for them.[11]

Because the deep rocks of our planet's interior are hot and molten, one would expect that gravitation would have produced an original surface of minimal topographic relief. Also, rain and wind inevitably wear down the land over time. Whether by erosion or equilibrium, the force of gravity should have produced and maintained a smooth two-dimensional surface. Instead, planet Earth has a dynamic two-tiered surface. Subduction at plate boundaries (forced by plate movement) lifts the continental platforms, maintaining higher land surfaces. But the critical reader may complain that, on a planet over 41 million feet in diameter, the 14,600-foot average difference between ocean depths and land surface is trivial. Not so. If the Earth's surface were smooth, the overall depth of water covering our planet would be about 8,800 feet (2,700 m). Simply stated: without plate tectonics we'd all be fish.[12]

## THE WET, BLUE AND WHITE PLANET

Finally, we come to the most extraordinary aspect of planet Earth: most of its surface is covered with water and much of its atmosphere is embellished by water vapor. We are the blue and white planet; nothing else in the solar system looks even remotely like Earth. Mercury, Venus, and Mars are drier than dust. Two of the moons of Jupiter may have liquid water beneath their solid frozen surfaces, but they are unlikely to support complex life-forms. Only Earth boasts a surface 70 percent of which is covered by the only liquid in the universe likely to sustain complex life-forms.

The scarcity of water on our sister planets suggests that Earth's oceans were the result of yet another grand accident. Planetary scientists generally agree that the early history of the solar system included a period of intense bombardment when planetary debris was scattered by the gravitation of the larger bodies. The pock-marked surfaces of both the Moon and Mercury bear witness to this violent early period. Even if the early Earth had a substantial amount of surface water, it is probable that much was lost during the Moon collision. Such an impact would have blanketed the Earth in thousand-degree rock dust that would have evaporated the ocean and destroyed whatever early forms of life might have developed. But then our planet got really lucky—again. Sometime after the Moon encounter and during the period of heavy large-object bombardment, we probably had yet another series of collisions, this time with a barrage of comets or "dirty snowballs."

Unlike the rocky inner planets and the gaseous giant planets, comets carry simple compounds and large amounts of water in the form of ice. As the recent Jupiter probe has suggested and the surfaces of our sister planets indicate, water isn't that common among the planets of our solar system. Armand Delsemme describes how the comets probably originated early in the history of the solar system, forming about five astronomical units from the Sun, near Jupiter, and beyond.[13] Thrown around by the gravitational effects of the giant planets over millions of years, the comets peppered the planets or ended up in the Oort Cloud. Although Earth's early minerals did contain significant amounts of water, there seems to be no reasonable alternative to the suggestion that our planet received additional water from the impact of cometary and asteroidal material. David Morrison suggests that, were it not for the rain of ice and carbon compounds, our planet might be as dry and lifeless as the Moon. "Life," he concludes "is a gift of the comets."[14]

The Earth's surface, three-fourths of which is covered by water and ice, provides not only vast oceans but also the humidity and cloud cover needed to spread rain over wide areas of land surface. Had our planet been covered with far larger land areas, it would be mostly desert. With smaller areas of land surface, Earth would have lacked large enough expanses to support a rich and diverse terrestrial flora and fauna. As luck would have it, our planet seems to have not only a generous supply of water, but also a very felicitous ratio of land surface to water surface. With water's unusual capacity to absorb and store heat energy, huge oceans also have helped to stabilize the Earth's climates as other factors have varied. And there's more. Our sister planets have either volcanoes (Mars) or a reconstructed surface (Venus), indicating they also have strong tectonic forces beneath their surfaces; but they have nothing similar to Earth's mobile plates. A number of earth scientists have suggested that a generous supply of water has helped "lubricate" the Earth's crust and made plate tectonics possible. Since hydrated rocks are usually more flexible and compressible than those lacking water, this suggestion makes sense. More important, water lowers the melting temperature of minerals under pressure, as salt on ice. Thus, a generous covering of water may be critical to having a dynamic mobile crust. In his recent survey of our solar system, Stuart Taylor writes: "Without water we would have no granite, no continents to stand on, few ore deposits, and no advanced technology."[15]

The oceans have been the mother of us all. It is difficult to imagine complex life-forms originating in any other medium. And here again, our dynamic crust may have played a significant role. Beyond the business of concentrating minerals, deep sea hot spots may have done something even more essential for our story. Protected from surface-vaporizing meteoritic bombardment deep on the cool ocean floor, volcanic geysers, with their bubbly mix of thermal energy and concentrated chemicals, may have been where early life-forms originated.

Not only does our planet have a generous and appropriate supply of water, our oceans seem to have been remarkably stable in regard to their chemical composition over much of the Earth's history. Why haven't the oceans become increasingly salty over time? With eroding rivers constantly bringing more dissolved salts to the ocean, the saltiness of the ocean should have been increasing continuously over time. Where did all the extra salt end up? One possibility brings us back to the thermal activity and recycling that take place at volcanic ocean ridges, subduction zones, and hot spots. It seems that ocean water percolating into deep fissures reemerges, after having been subjected to many chemical reactions, enriched in minerals but depleted in salt. Perhaps these processes helped our oceans to maintain a stable chemical environment over thousands of millions of years.

## A THIN ATMOSPHERIC BLANKET

Unlike the ocean, the Earth's atmosphere has changed profoundly over time. Extremely low concentrations of neon and argon in today's atmosphere indicate that the original mixture of gases of the solar nebula was lost early in Earth's history. (Because Earth's gravitational field is relatively weak, helium and hydrogen were quickly lost to outer space.) After the Moon collision, outgasing from volcanic activity probably created a renewed atmosphere rich in carbon dioxide, methane, and ammonia. A good thing too. If astronomers are correct in their calculations, a young Sunlike star gives off 25 percent less energy in its early life than during its long maturity. A cooler Sun should have resulted in a frozen early Earth—which obviously didn't happen. Internal heat from a higher concentration of radioactive elements, as well as bombardment by interplanetary debris, would have warmed our planet in its earliest years. Also, a thicker atmosphere, with carbon dioxide perhaps a thousand times denser than it is today, helped keep the early planet warm. Though vastly different in densities, the present-day atmospheres of both Venus and Mars are about 95 percent carbon dioxide, as compared with 0.03 percent carbon dioxide on Earth today. Water and biology have made the difference. Heavy rains would bring carbon dioxide into the sea as carbonic acid, slowly building huge deposits of carbonates. In addition, microscopic sea creatures extracted carbon dioxide from sea water, building tiny carbonate domiciles, and, over time, forming massive marine deposits. With carbon dioxide being incorporated into a variety of organic materials and becoming sequestered in geological deposits, its concentration in the atmosphere slowly declined, allowing more of the Earth's heat to escape into space. After bacteria had developed photosynthesis they began expelling free oxygen into the ocean and atmosphere. Free oxygen would react with a great variety of compounds, including methane and ammonia. This helped remove high levels of those gases from the atmosphere. How convenient and how fortunate! Just as living things were reducing the high concentrations of heat-trapping carbon dioxide, methane, and ammonia in the atmosphere, the adolescent Sun began warming up. This delicate balance between the changing composition of our planet's atmosphere and a slowly warming star may have been another big lucky break in the earthly saga.

Today's atmosphere, rich in ozone-generating oxygen, together with the Earth's magnetosphere, protects us from the Sun's ultraviolet radiation, X rays, and gamma rays produced by solar flares, as well as high-energy radiation from elsewhere in the galaxy. The atmosphere also keeps us warm. According to calculations based only on the Sun's energy output and the Earth's reflective characteristics, our planet's average temperature ought to be about –5°F (–20°C), well below the freezing point of water. The reason we have the comfortable climate we have is because we are encased within a thin but efficient heat-conserving atmospheric blanket. How thin our aerial blanket is can be illustrated by the fact that if the Earth were five feet in diameter, the breathable lower atmosphere would be less than a tenth of an inch thick. Nevertheless, and for over 3,000 million years, the Earth's thin atmosphere has helped us avoid frying in a runaway greenhouse calamity or slipping into an ever-colder glaciation.[16]

While early life may have had beneficial effects for our atmosphere, some of these effects were also deadly. Photosynthesis, capable of splitting apart the water molecule, created the first major case of worldwide pollution with a dangerous poison: oxygen. We can't live without oxygen, but for early and primitive forms of life, this highly reactive element was life-threatening. Oxygen is what keeps fires going and oxidizes (rusts) so many substances. As photosynthetic life expanded, living things had to develop ways of protecting themselves in an environment of increasing oxygen pressure. An atmosphere rich in oxygen was both the product of living activity and, as we shall see in chapter 4, an elixir that helped give rise to vigorous new forms of life. Since that time, and over the last 500 million years, oxygen concentration in our atmosphere appears to have been remarkably stable. Had its concentration dropped below 5 percent of the atmosphere, complex animals would have been unlikely to survive. Had it exceeded 35 percent, terrestrial vegetation would have been subject to incessant fire. An atmosphere made up of 21 percent oxygen (and 78 percent innocuous nitrogen) is another fortunate feature of our terrestrial environment. Created by the activities of life itself, no other planet or satellite in the solar system sustains an atmosphere so richly endowed with the element that helps power our every thought: oxygen.

Later, an escalating exuberance of life would further change the atmosphere, as land vegetation developed and prospered. The great coal forests of 300 million years ago apparently drew down levels of carbon dioxide to such an extent that a long period of ice ages followed.[17] Likewise, the success and diversification of the flowering plants in more recent times may be responsible for similar cooling effects.[18] Now, after two centuries of busily burning fossil fuels, we humans are elevating carbon dioxide concentrations once again.

Dramatic discoveries by recent space probes visiting the planets and their many moons have revolutionized our understanding of the solar system. These new findings have shown astronomers how history can make as huge a difference in the life of celestial objects as it has in biology and human affairs. Each planet and every moon in our solar system has been on its own special odyssey. Earth and Venus, once considered "sister planets," have turned out to be profoundly different. Likewise, each of the "hot Jupiters," detected at nearby Sunlike stars, also appears to be the product of unique historical circumstances. All these observations illustrate how slight differences in initial conditions and later accidental encounters can lead to an enormous variety of outcomes. Because of biological activity based on the photosynthetic splitting of water, Earth's atmosphere became uniquely rich in oxygen. Thus, a combination of lucky breaks and unusual trajectories confronts us with the very real possibility that our blue and white planet may be unique, not only in its own solar system, but among many thousands of other solar systems as well.[19]

We have been extraordinarily lucky. Earth has a fine location relative to the Sun, a zesty spin with a bit of a tilt to spread annual weather patterns around, a dynamic two-tiered surface, and lots of liquid water. Two lucky accidents in the Earth's early history may have played important roles in making the Earth so hospitable. Collision with the proto-Moon allowed for an entirely new atmosphere to form, gave us greater mass, a quick spin, and a stabilizing companion. Later collisions with abundant cometary material brought us an additional dose of life-sustaining water. Only on such a stage could the panorama of life achieve an exuberance of bacterial diversity, of coral reefs teeming with fish, of the majesty and diversity of rain forests, and of a species contemplating the stars.

We're not just talking Goldilocks's orbit here; we've got the Goldilocks *planet*. Our gravity, spin, crust, atmosphere, and water are all "just right." And we've barely begun our journey; we'll see how a lot of other lucky breaks were required before one species of earthlings began building radio telescopes to scan the heavens. The early evolution of life, however, progressed very slowly, as tiny living things achieved greater internal complexity. We will next survey the time before complex animals make their first appearance, a period during which our planet would circle its star 4,000 million times.

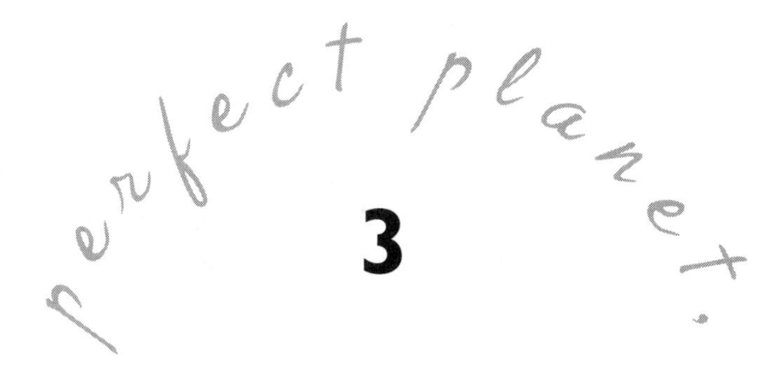


# 3

# THE FIRST 4,000 MILLION YEARS

clever species

We contemporary humans are the latest links on a chain of being that reaches back to the origin of life on our planet. No matter how complex we fancy ourselves to be, each of us began as a single fertilized cell. Such cells are themselves very complex and required thousands of millions of years to evolve. We should understand these early stages in our history when discussing the probabilities of our uniqueness. Basic cell biology, the genetic code, and even sex were fundamental to the further development of more complex living things. And we will need to review these aspects of the living world that began early in our planet's history.

Today's biological diversity, ranging from lush prairies to humid rain forests and an ocean full of swimming creatures, developed only in the more recent phases of Earth's history. Fossils of recognizable larger living organisms are not known from before about 560 million years ago. Four thousand million years of geological deposition contain little physical evidence of living things. Nevertheless, those early times were important. Life itself had come into being and the bacteria diversified. With time, and though still minuscule, larger cells achieved new levels of internal complexity. Such cells made possible even greater structural diversity. As in the case of so many other historical sequences, the early stages of our epic proceeded at a glacial pace and then became more rapid with each succeeding era.

## IN THE BEGINNING

The earliest age for materials within meteorites, based on evaluation of isotopic ratios, is about 4,560 million years ago. This is probably the time when Earth was becoming an accreted planet. The ensuing "Hadean period" lasted from that time to somewhat less than 4,000 million years ago. The name is appropriate; it was a time of chaos in the solar system. Crater-marked surfaces of Mercury and much of the Moon bear witness to this period of intense bombardment. Because this period extends for more than 500 million years *after* the date of the oldest meteoritic rocks, it is called the "*late* heavy bombardment." Planetary scientists theorize that this period, when rocks and comets were falling from the sky with explosive regularity, extended up to about 3,800 million years ago. That was the bad news. The good news was that these extraterrestrial missiles may have added not only water but significant amounts of phosphorus and other critical elements to the earth's crust.

Despite continued sporadic bombardment and occasional vaporization of ocean surfaces, the deep sea may have provided a special environment in which the earliest forms of life could assemble themselves. Interestingly, living species able to flourish in extremely hot water are found in all the more primitive lineages of living bacteria—a fact that fits nicely into scenarios of life beginning in a hot aqueous environment. Today, there are a few groups of bacteria that live close to the boiling point of water under high pressures, such as those found near volcanic vents and fissures in the deep sea. It is not clear whether these "hyperthermophiles" are highly specialized for these unique habitats or are still living in the same kind of hot, oxygen-free setting they enjoyed very long ago. Among these heat-loving bacteria is a group that is now being recognized as something quite distinctive—the Archaea.

Studies of bacterial genes indicate that what we had been calling "the bacteria" contain at least two major and distinctive groups. One group contains most of the bacteria we had been familiar with, but the other includes strange forms found in very hot waters. Today we call this latter group the Archaebacteria, or Archaea for short. Strong evidence for the claim that Archaea are quite different from other bacteria (Eubacteria) comes from sequencing all the genetic information (called the genome) of the Archaean species called *Methanococcus jannaschii*.[1] This microbe was first discovered at a hot spot deep on the floor of the Pacific Ocean, where it lives at temperatures between 120 and 200°F (50–95°C) under intense (200 atm) pressure. Utilizing carbon dioxide, nitrogen, and hydrogen, these unusual bacteria obtain their livelihood from reactions reducing carbon to methane; moreover, they can function only in an environment that has extremely low levels of free oxygen. Such a sensitivity to oxygen, together with their high temperature environment, makes Archaebacteria a model analog for a very early form of life in today's world. Of the 1,738 genes that have been identified in this bacteria's genome, a surprising 56 percent have no analog in other living organisms. Most of its genes regulating metabolism, cell division, and energy production are similar to those already known in other bacteria. In contrast, most of its genes involved with gene processing are similar to those found in organisms whose cells have a nucleus. Such unusual gene relationships suggest (1) that the Archaea represent the living descendants of a very ancient group, and (2) that these early groups were able to exchange important genetic information among themselves. Recent research indicates that bacteria can exchange genetic information between widely different lineages. As we learn more about these ancient organisms, our image of a "tree of life" is beginning to look more like a trellis.[2]

Just how life began on the oxygen-free surface of a young Earth is a subject of much speculation. Some astronomers think that a cool young Sun may have shone down on an Earth where ocean surfaces were frozen solid, though subject to frequent vaporization from intense asteroidal bombardment—not a likely environment for the origin of life. In contrast, the deep ocean floor, with tectonic rifting producing continuous eruption of lava and hot gases in an aqueous high-pressure environment, may have produced conditions ideal for the origin of life. But the leap from complex organic molecules to life-enhancing enzymes and a chemical system of information transference is huge.

How long-chain DNA and RNA molecules came into being and how such molecules first created complex proteins are questions that fascinate biochemists and inspire a continuing search for testable scenarios. (DNA is deoxyribonucleic acid; RNA is ribonucleic acid. Both are long-chain polymers that carry genetic information in living things.) Many scientists think that the living world may have begun with RNA. In the contemporary world, RNA is central to protein synthesis, carries information, edits transmitted information, and can itself act as a catalyst or enzyme. After an early "RNA world," the next step may have been the origin of protein-synthesizing ribosomes, with DNA becoming the primary repository of hereditary information only later. It's important to keep in mind that DNA cannot be replicated without a complex of associated enzymes in just the right cellular environment. Moreover, the leap from chemistry to the intricacy of even the simplest bacterial forms of life is enormous. For many scholars, the creation of living, reproducing beings out of a rich mix of organic chemicals by natural processes was one of the most difficult *and unlikely* steps in the history of life on our planet. But perhaps this was not as difficult a process as it at first appears to be.

## LIFE: ORDER OUT OF CHAOS

Biophysicist Stuart Kauffman has argued that spontaneous order is enormously more pervasive throughout the universe than anyone has imagined. He believes that in a rich mix of molecular compounds, self-sustaining networks of reactions will arise spontaneously, requiring only that "extremely complex webs of interacting elements are sparsely coupled." Kauffman often employs phrases such as "mathematical order in complexity" and "between order and the edge of chaos."[3] He suggests that there are subtle rules of inanimate organization at higher levels of organization in the microscopic world around us.

Living in a seasonally frigid environment, I like to visit the edges of small brooks just after temperatures have plummeted well below freezing. As the air becomes very cold, ice begins to develop around the brook's edge, slowly building outward over the surface of the rushing water. The forms and patterns of the newly formed ice and the variety of patterns of trapped air within these ice formations can be surprisingly intricate. Some patterns in the ice recur often, but the variety of formations for so simple a system is impressive. At a smaller scale and despite the fact that they are always relatively flat and hexagonal, the great variety of forms exhibited by snowflakes challenges credulity.[4] Surely, similar processes are operative at submicroscopic molecular scales. Self-organization of thin molecular films, gels, and complex membranes are examples that have been suggested as important in the beginnings of life. Even something as mundane as clay has been proposed as playing a significant role in life's origin. Organic molecules oriented on an electrically charged, two-dimensional surface have much less freedom than in a hot soup. With unusual electrical properties on its flat, microscopic, crystal surfaces, clay may have facilitated molecular orientation and coupling—critical early stages in building more complex molecular aggregations.

Evidence of past life on the planet Mars would greatly bolster the argument that simple life is highly likely to develop under the right conditions. Along these lines, minute spherules containing carbon found in a Martian meteorite were recently cited as evidence for life.[5] With such a bold claim, our National Science Foundation and space agency quickly made 2.3 million dollars available for the further study of this particular meteorite. (Try to get that kind of sudden generosity to study a threatened rain forest.) By late 1998 these further studies had concluded that there is *no convincing evidence* for life in this particular bit of Mars. So let's get back to our subject: life on planet Earth.[6]

Philosophers, biologists, and many popular authors have imagined the origin of simple, bacterialike life-forms to be one of the most difficult steps in the history of life. But if people like Kauffman are correct, this may not be the case at all. Rather, given the right conditions, life may be the naturally emergent expression of subtle but universal rules of order. "Life is the natural expression of complex matter. It is a very deep property of chemistry and catalysis . . . ," claims Kauffman.[7] Mathematician Per Bak believes that complex systems can spontaneously form a dynamic state he calls "self-organized criticality," and that this is nature's way of making large transformations over short time scales.[8] Biochemist Christian de

Duve states that "Life is an obligatory manifestation of the combinatorial properties of matter . . . bound to arise under the [proper] prevailing conditions. . . ."[9] Also, if we imagine the creation of life as the accidental result of a felicitous natural experiment, it's worth remembering that the "laboratories" in which these many experiments would occur are smaller than the tip of a needle. Along deep ocean fissures there would have been countless potential sites for such microscopic experimentation—continuously available over many millions of years. Under such circumstances, the almost impossible would have become highly probable soon enough.

An environment supporting living things requires the continuous flux of the right kind of energy through the system; on this lovely planet we happen to have two such energy sources. For the prolific evolution of life, solar radiation cascading over the surface of the planet has been our primary energy source. However, for the origin of life, it may have been the continual outpouring of hydrothermal energy in a mineral-rich environment of cool deep-sea water that helped get things started. Far from the high-energy irradiation near the ocean's surface, and insulated from periodic asteroidal bombardment, volcanic vents along the ocean floor may have been the environment where life began.[10]

## THE EARLIEST EVIDENCE OF LIFE

Evidence for the earliest forms of life has always been the subject of some controversy. There may be little hope of finding fossilized bacteria in the oldest rocks; thousands of millions of years of geological metamorphoses are likely to have destroyed their minute morphological features. However, the earliest forms of life may have left a chemical signature in the rocks. Biochemical processes bring the isotopes of carbon together in percentages that are different from those of other natural physical and chemical processes. Using sophisticated modern microanalysis, chemists have been able to derive meaningful data from minute carbon inclusions in ancient mineral grains. A recent report claims that elemental carbon trapped in 3,850-million-year-old apatite has the isotopic attributes of organic carbon produced by biological processes.[11] Because they contain a higher percentage of the carbon 12 isotope than are produced by other processes, a biological origin is implied.

Some early life-forms probably obtained their energy by reducing ferric iron to ferrous iron. Although rare today, this simple reaction could have supported life activities for hundreds of millions of years—before oxygen, nitrates, and sulfates became abundant on our planet. However, by "inventing" photosynthesis, bacteria with complex light-absorbing pigments took a giant leap forward in the history of life. "Probably the most important evolutionary innovation on the Earth, if not in the solar system and the galaxy, was photosynthesis," declare Lynn Margulis and Karlene Schwartz.[12] Through photosynthesis, the energy of sunlight is transformed into the calories of food molecules that energize the vast majority of life activities. Once this reliable energy source had become part of microbial life, evolution could begin to run uphill, countering the universal tendency of things to fall apart and dissipate—the second law of thermodynamics. With photosynthesis, life on our planet had the firepower to proliferate.

Because bacteria are so small and do not have resistant walls, they and other early life-forms are very rarely found as fossils. Luckily, there are situations where photosynthetic bacteria form thin surface films that trap minute sediment and deposit carbonates, creating stromatolites. They are rare today but were common in the early fossil record.[13] Living in shallow marine embayments, stromatolites are rounded, cushionlike, club-shaped or columnar rocky structures, generally one to two feet (30–60 cm) in diameter and a few inches to six feet (10–200 cm) high (see Figure 1 on page 86). What makes them distinctive is that, in addition to their rounded, three-dimensional shape, they have a thin, repeating, internal layered structure. In the living examples, microscopically thin layers of photosynthetic bacteria or algae cover the surface of the mound where they trap sand and debris. As they continue to grow, they form new layers, thus creating an expanding laminated cross section. The earliest stromatolite-like structures occur in strata dated at about 3,500 million years old, only 300 million years after the end of the late heavy bombardment. A few have the preserved fossil remains of minute bacterialike forms, but most do not. Were all these ancient stromatolites created by the action of thin bacterial films?

Many geological processes can give rise to laminated structures. Conelike stalagmites often have a layered structure due to chemical changes in the water from which they are formed. The water's wave action can often produce thin-layered strata of differentiated sediments. However, the unusual three-dimensional shape of both fossil and living stromatolites suggests that they are not the products of simple physical factors. All of the living stromatolites are built by microscopic algae or bacteria. But they are rarely encountered today, found only in very warm, highly saline water where grazing animals are absent. Before grazing animals evolved, stromatolites were common along shallow sea shores; their fossils provide evidence of active life over more than 3,000 million years of Earth's history.

Bacteria, both those that can photosynthesize and the vast numbers that cannot, are, in many respects, the world's most successful life-form. Some are found in superheated water at high pressures at the bottom of the sea, and even in hot springs close to the boiling point of water. Some live within microscopic spaces in rocks deep within the Earth where they use chemical reactions to sustain themselves. Many live in environments devoid of oxygen where higher forms of life cannot survive. They live everywhere on our planet where life can exist, and exhibit a greater variety of biosynthetic pathways than any other group. Some bacteria are essential to the proper functioning of our intestinal tract, while others can kill us with their toxins. Bacteria are so widely distributed and so common that, despite their microscopic size, a few biologists speculate that their total volume exceeds the volume of all other living things on the planet.

Life's first major development, the bacterial stage, came relatively quickly. This was a period in which cells were small and lacked a nucleus. The following stages of larger, more complex cells required far more time before they'd make their appearance in the fossil record. It is not until 1,500 million years after the stromatolites made their debut that we find evidence for larger, more complex forms of life. Filamentous algaelike fossils have been identified in rocks dated at about 2,000 million years old in Michigan's Upper Peninsula.[14] These thin chains of fossil cells are almost ten times thicker than bacterial filaments, similar to the differences found between algal and most bacterial filaments today. While some geologists refer to the period between 2,000 and 1,000 million years ago as "the dullest period in Earth's history," this may have been the time when simple life forms were gradually achieving greater complexity.

## THE NUCLEATED (EUKARYOTIC) CELL

Cells with a nucleus were a major advance beyond the structural simplicity of the bacterial cell. All members of the plant, animal, and fungal kingdoms have nucleated (eukaryotic) cells. Yeast is a nucleated cell and one of our favorite microscopic organisms; we've been using it for thousands of years to make bread rise and to produce alcoholic beverages. A single-celled member of the fungal kingdom, its entire hereditary information content has recently been published. The 12-million base-pair sequence was the first described for any organism with nucleated cells and those many base pairs are estimated to encode about 6,200 genes. The yeast genome is over six times larger than the first completely sequenced bacterial genome. But the sixfold difference in hereditary information is only part of the story. Bacteria and their kin differ in profound ways from most of the living things we are familiar with. To begin with, bacterial cells are on average *one-tenth* the diameter and have *one-thousandth* the volume of average cells in higher organisms. (An exception is the photosynthesizing blue-green bacteria that have much larger cells than their nonphotosynthesizing bacterial cousins.)

A major distinction of bacteria is that their solitary cell lacks a nucleus, the spherical organelle in which chromosomes are confined in eukaryotic cells. In bacteria, genetic information carried by the DNA resides freely as a single long, thin loop within the cell. A significant measure of the difference in complexity between bacterial and nucleated cells is the time it takes an individual cell to reproduce. Given sufficient resources, the average bacterium can replicate every twenty to thirty minutes; the average nucleated cell takes about twenty hours. More information and increased complexity simply take more time. In turn, complexity within the nucleated cell makes possible building more elaborate forms of life.

The nucleus-bearing cell has a complex internal organization with an almost invisible cytoskeleton and a variety of organelles. (The cytoskeleton provides an internal architecture for the cell, while organelles are small functional structures within the cell). The aqueous material throughout the cell but outside of the nucleus is called the cytoplasm, and here is where most of the cell's life-giving activities are carried out. Energy for the cell from oxygen-consuming respiration takes place within dozens to thousands of little lozenge-shaped organelles called mitochondria (singular *mitochondrion*). Their numbers depend on how active the cell is; muscle and liver cells have thousands. While plant cells also have mitochondria, they usually harbor another class of special organelles as well: chloroplasts that capture the energy of sunlight and begin the process of building energy-rich molecules. In addition, all living cells contain ribosomes, minute organelles that build amino acid chains. These chains, in turn, are transferred to the interior of tubelike parts of the endoplasmic reticulum (a membranous entity within the cytoplasm) where they become folded into the specific three-dimensional protein that will do the work of the cell. By continuously forming and dissolving themselves, the elements of the cell interior are able to bring together and sustain those many processes that keep the cell alive.[15]

The nucleus is usually spherical and bounded by a thin membrane. The chromosomes are confined within this membrane, and here is where chromosome duplication and transcription occur. "Instructions" transcribed from the DNA of the chromosomes are carried out of the nucleus by messenger RNA to the ribosomes. What the nuclear membrane does is to *confine* the process of making copies of the genetic code (replication) and the process of putting this information onto RNA (transcription) within the interior of the nucleus, before sending it out into the cytoplasm to do its work. There the ribosomes will use the information from messenger RNA (in the process of translation) to create the peptide chains that will become the complex proteins doing the real work of the cell.

Just as the nucleus is different from anything found in bacteria, the nature of the outer cell wall in higher organisms also differs significantly from the walls of bacteria. Enclosing the entire nucleated cell, the outer cell membrane is quite complex with special surface areas through which foreign substances can be ingested and waste products can be excreted. Many single-celled nucleated organisms have the ability to engulf food particles within the cell where they can be digested without loss of digestive enzymes. Bacteria must exude digestive enzymes through their walls, then reabsorb the enzymes and "digested" nutrients—not a very efficient way to get dinner.

Surely, the most critical part of the nucleated cell's life cycle is the preparation for and execution of cell division, as one cell gives rise to two daughter cells. The process begins during what appears to be a quiescent period of the cell's activity, when the chromosomes are thin, fully extended in length, and virtually invisible within the nucleus. During this "interphase," as the cell is carrying on its regular life activities, information from the chromosome is transferred to the cytoplasm as needed. Toward the end of this interphase period the chromosomes will have produced sister strands, and they will be ready to condense into the thickened structures that inaugurate cell division. As they contract and thicken, the nuclear membrane dissolves, and the now clearly visible chromosomes become aligned along a flat plane at the center of the cell. Two sets of spindle fibers develop and become oriented facing opposite sides of the cell. Meanwhile, the sister strands of each chromosome remain attached at a site having two attachment points, the kinetochores. These two kinetochores then orient themselves on the division plane, divide to separate the sister strands, and begin moving along the spindle fibers in *opposite* directions, dragging the chromosome strands along behind them. In this way the chromosome strands are separated and move to opposite sides of the cell; each daughter cell receives exactly the same chromosomal complement as the original cell. Finally, the spindle fibers dissociate, new nuclear membranes form to enclose the chromsomes, and a plate forms in the same plane on which the chromosomes were earlier aligned. Here is where a new cell wall forms in plants, or where the membrane of the animal cell pinches in two, producing two cells where there had only been one. Carefully cadenced, the cycle of cell duplication (called mitosis) consists of a series of subtly coordinated changes in the cell's internal environment.[16] More impressive still is the realization that *all* complex living organisms develop from a single cell by many generations of such cell divisions.

## CREATING THE COMPLEX CELL

Not only is a nucleated cell much larger than a bacterial cell, it is far more intricate and contains a number of elaborate organelles. The most important of these are the nucleus, ribosomes, mitochondria, endoplasmic reticulum, and, in plants, the chloroplasts where photosynthesis takes place. (With the exception of ribosomes, bacteria lack these structures.)

Mitochondria are the energy producers of the nucleated cell. Here is where glucose (from photosynthesis) is broken down into carbon dioxide and energy-rich hydrogen atoms. The hydrogen atoms, in turn, cascade through a series of cytochrome enzymes to be united with the oxygen we breathe. The first part of this process produces carbon dioxide and does not require oxygen but captures only 2 percent of the energy available in overall glucose breakdown. The second stage, where cytochrome enzymes transfer hydrogen atoms to oxygen, captures 39 percent of the potential energy—producing water as its "waste." The remaining potential energy is dissipated during this cascading dance of chemical interactions and during the process whereby energy is transferred to adenosine triphosphate (ATP)—the molecule that dispenses energy wherever it is needed within the cell. Using oxygen in the second stage of respiration gives us most of the energy we need to live our fast-paced lives. Many simple forms of life that do not employ oxygen-consuming respiration can thrive without oxygen, but theirs is a much slower-paced life of modest energy consumption.

Interestingly, mitochondria are similar to many bacteria in their general dimensions and shape. Also, the membranes lining the inner recesses of the mitochondrion resemble the membranes of bacteria, rather than the membranes of the animal or plant cells in which they are found. More important, it was discovered that mitochondria carry some of their own special hereditary information in a circular bacteria-like DNA strand. Thus the mitochondria are self-replicating little bodies within the cytoplasm of the much larger nucleated cell. Chloroplasts of plants, much larger than mitochondria, also carry some of their own DNA blueprints, quite separate from the DNA sequestered in the plant cell's nucleus.

As early as the late 1800s, a few scientists had speculated that certain cell organelles may have had a bacterial origin. This idea was not pursued seriously until the late 1960s when it was elaborated and championed by Lynn Margulis. She proposed that nucleated cells had achieved some of their complexity by *incorporating bacteria* into their cellular contents over evolutionary time.[17] Her hypothesis claimed that mitochondria and chloroplasts had become elements of the larger nucleated cell by forming a close symbiotic partnership followed by complete union. (This is now called "endosymbiosis." Symbiosis is a close positive relationship between two different organisms, and *endo* refers to the interior of the cell.) This hypothesis has since been strongly supported by detailed analyses of the respiratory cytochrome enzymes in mitochondria, which have many similarities with purple nonsulfur bacteria. Likewise, chloroplasts of green plants share many molecular similarities with living blue-green cyanobacteria. It now seems clear that, beginning in closely symbiotic association over millions of years, smaller bacterial partners finally became thoroughly integrated parts of the machinery within the larger eukaryotic cell. The endosymbiotic hypothesis makes clear why mitochondria and chloroplasts still carry a bit of their own special hereditary information; they were once independent beings.

More significantly, the endosymbiotic hypothesis helps us answer the Reverend William Paley's challenge: when you find something as complex as a watch, with its carefully crafted parts, you know there has to be a watchmaker. Paley used this argument early in the nineteenth century as evidence for a divinely created natural world.[18] Indeed, the complexity of the living eukaryotic cell presented science with this problem: how might such a complex cell come into being *without* providential design? The endosymbiotic hypothesis helps us understand how "chance," operating over a thousand million years, might have put together something as intricate as the plant or animal cell. In fact, recent research has shown that other endosymbiotic fusions have taken place independently in several different lineages.[19] Many people may feel uncomfortable with an evolutionary theory based on "random" events; but picking up and incorporating a bacterial symbiont or accidentally duplicating an important gene and then elaborating the "extra gene" for new roles are quite complex and progressive steps. They are random only in the sense that they originated accidentally.

The complex nucleus-bearing cell is a power-hungry device; think about the dynamics of chromosome condensation, orientation, and separation during cell division. Not until energy-producing mitochondria became fully integrated was it likely that nucleated cells could do the demanding work of moving larger chromosomes with their baggage of increased genetic information.[20] The acquisition of fully symbiotic mitochondria may have been the earliest of several critical innovations giving rise to larger and more complex cells.[21] Other steps in building the more complex nucleated cell are indicated by the fact that our own cells, as well as those of much simpler organisms, carry genes that are very similar to genes found in Archaebacteria *in addition to* genes similar to those of bacteria. Clearly, many different lineages took part in building more complex cells.

In today's world, symbiotic bacteria play an essential role in helping animals as disparate as cows and termites make their living. Animals lack the enzymatic machinery that can tear nutritious glucose molecules out of cellulose fibers. Without their bacterial symbionts, cattle could extract very little nutrition from straw and termites would starve eating wood. Bacteria also play an important role in our own intestines, helping us digest our food and, perhaps, supplying us with a few vitamins we cannot synthesize ourselves. In the world of plants, legumes and a few others host nitrogen-fixing bacteria in special nodules on their roots. This symbiosis provides these plants with an extra source of nitrogen. Unlike the early symbioses postulated by Margulis, these modern bacterial symbionts live within their hosts but not within ordinary cells.[22] Having delved into cell complexity, let's get back to the business of creating new cells, which carry genetic information from generation to generation.

## THE HEREDITARY INFORMATION

Regardless of whether the living specimen is a minute bacterium, a larger yeast cell, or a complex multicellular plant or animal, the basic hereditary information for all living things is based on double-stranded DNA and its close partner, single-stranded RNA. Deoxyribonucleic acid (DNA) is a long-chain compound made of three basic components: a five-carbon sugar (ribose), a phosphate group, and one of four different nitrogen-containing bases: adenine, cytosine, guanine, and thymine (in RNA, uracil takes the place of thymine). Each single strand is made up of sugar molecules connected to each other by phosphate groups in a very long unbranched chain (a polymer). One nitrogenous base is attached to each sugar molecule and projects outward from the axis of the chain. The crucial aspect of this long, single-axis molecule is that the laterally projecting nitrogenous bases can form weak hydrogen bonds with complementary bases on an *adjoining* parallel DNA chain. Only the cytosine-guanine (C-G) and adenine-thymine (A-T) pairings have the right size and form to keep the two strands of DNA at the proper distance and hanging together. Better than that, these same pairings have the conformation that, when the strands are separated, can specify the building of new and identical complementary strands. In this way, each strand becomes the template for building a new complementary strand, while the long sequences of four differing bases along the length of the chain carry the information in a linear sequence. Thus, the long, thin double-DNA chain resembles a spiraling ladder in which the "rungs" are the paired bases loosely held together by weak hydrogen bonds. These rungs can be separated at the center to split the ladder longitudinally (like a zipper), with each half-ladder specifying the construction of a new other half. The special characteristic of the long, thin chains of DNA and RNA is their ability to "store information" in linear form (the long sequence of bases) that can be easily copied, thanks to the strict base-pair complementarity (C-G or G-C and A-T or T-A).

Whether flower, fish, or fowl, DNA's laterally projecting bases are the "letters" forming the "words" or genes in the hereditary instruction book that make living things what they are. These words (genes) are arranged along the length of the threadlike DNA chain that is the essence of the chromosome. Unlike phonetic human script with twenty to two hundred letters, the DNA code really has only four: the A-T, T-A, C-G, and G-C couplings. But the working parts of the cell are mostly complex proteins, themselves constructed out of as many as twenty different amino acids. How do we get from a string of base pairs on the DNA chain to a string of amino acids that can become part of a complex protein? What's needed is a code for translation. Complementing the discovery that DNA is a universal carrier of hereditary information, biologists soon learned that all living things share the same three-base code to designate the twenty different specific amino acids. The base-pair triplets (codons) have some redundancy, with the more common amino acids having several synonymous codons; but from bacteria to barracudas, or algae to redwoods, all use similar coding. In this way information carried by nucleic acid chains is "translated" to form the amino acid chains that become the long, folded proteins that do the work of the cell. As a consequence, and unlike words in human languages, which are usually limited to fewer than fifteen letters, genes carrying the coding for long, folded protein molecules can be hundreds to thousands of base pairs long. Remember, it is the proteins that form the enzymes and the structural elements of the cell doing the actual work of the living cell; genes do no more than carry the hereditary information code. And they *alone* carry that information; there is no way for the proteins or other elements of the cell to add information to the hereditary code.

Luckily, just as many words can be understood if they are slightly misspelled or mispronounced, most genes have base pairs that can be changed without destroying the functioning of the resultant gene product. When such changes occur, they are called neutral mutations. Likewise, there are elements of the gene that cannot be tampered with, where the wrong base pairs render the gene nonfunctional. For example, there is a histone gene in which all but two of the 136 base pairs are exactly the same in yeast and in mammals, two lineages having been separated for perhaps as much as a thousand million years. Because histones are a central element in chromosome structure, those cells with defective copies of the histone gene simply cannot replicate.

While genes might be just as efficient swimming around the cytoplasm by themselves, the reason for their being strung together on chromosomes is so that they can be carefully duplicated and equally partitioned during cell division. This enables them to carry information forward from generation to generation.

The problem of duplicating the genetic code over and over again presented early living things with the challenge of quality control. In recent years the cellular mechanisms for gene repair and gene-splicing have become more clear. Replication of DNA produces about three errors per 100,000 base pairs. However, replication is followed by "proofreading enzymes" that can clip out most errors, and the error rate drops to about one error in 1,000 million base pairs. This ability of plant and animal genomes to make copies of themselves with so few errors, despite duplicating millions of base pairs, is an amazing feat. Of course, if they couldn't maintain high replicating accuracy, complex multicellular life-forms simply wouldn't be here. On the other hand, DNA replication, like just about everything else we know of, is subject to "Murphy's Law" (*What can go wrong, will go wrong*) and rare errors get through.[23] Though the vast majority of such errors are detrimental and swept away by an unforgiving environment, the minute fraction adding viable variations to the population are the basis for continuing adaptation to environmental change and for evolutionary innovation.

The genes of bacteria were the first to be studied in detail, and they were found to occur in neat order along the single circular chromosome. It was expected that nucleated cells would have similarly simple chromosome structure, but ongoing research has shown that the genome of nucleated organisms is far more complex. Unlike the case in bacteria, there are both coding regions (exons) and noncoding regions (introns) within genes having an overall coding function. The region corresponding to an intron is usually spliced out of the RNA copy after transcription. In addition to splicing, there are several types of messenger RNA editing, allowing a single gene to be modified to produce more than one kind of protein. These subtle ways of transforming the DNA sequence into a variety of RNA derivatives allows for a great variety of controls and variations. The simplistic notion that genes in the DNA of a chromosome are like words in a text may have been sufficient to understand bacterial genetics, but it is inadequate to describe the genetics of more complex nucleated organisms. The "miracle" of creating a bird or a butterfly from a DNA blueprint apparently involves more than just the blueprint. Messenger RNA editing, multiple interactions of modifier and regulatory genes, developmental programs controlled by critically timed protocols, and the emergent properties of developmental complexity itself all contribute to creating the variety of living creatures inhabiting our planet. We are not the simple products of our genes but the complex creation of many interacting factors.[24]

We should be careful that all this talk about genes doesn't create the impression that there is a simple one-gene–one-function or one-gene–one-trait relationship for most of the features that characterize a complex plant or animal. Gregor Mendel's genius was to have searched his monastery garden and to have carefully tabulated traits that were, indeed, determined by one or two factors (genes), and to discover that these factors passed *unchanged* through several generations. However, a huge majority of traits are the product of many genes and of interactions with other genes—and with the environment. Simple genetic reductionism—one gene, one trait—is popular because it does account for some nasty hereditary diseases; unfortunately, most of biology is not so simple.[25]

For many complex organisms the unraveled DNA strands from each chromosome laid out in a straight line end to end would extend two to four feet in length. Thanks to the fact that these chains are so submicroscopically slender and carefully coiled on tiny spools, they can be accommodated within the nucleus. The amount of total DNA in a single cell can vary greatly in different groups and even in related organisms. For example, the genetic code of the rice plant is about one-tenth the length of the barley plant's genetic code, despite the fact that they are both grasses. Nevertheless, these two important grain species appear to contain about the same number of functional genes. The difference in the amount of their DNA is largely made up of repetitive DNA lacking a clear hereditary function. As in the grasses, there may be large differences in DNA content among different animal lineages; the lungfish has thirty-six times as many base pairs in its genome as we humans do—and they're supposed to be a lineage of living fossils, dating back over 300 million years.

Recent estimates suggest that we humans have about 35,000 genes among the 3,000 million base pairs in our twenty-three chromosomes. Individually, our genomes are about 99.9 percent alike, which means that we differ by about three million base-pairs—no wonder each of us is as unique as we are. Genes make us blond or brunette, male or female, and, together with environmental variables, determine whether we will be tall or short, smart or not so smart. More fundamentally, our hereditary makeup determines that we will have two arms, two ears, one nose, and all the many other general characteristics of human beings. When you think about it, our genome has an extraordinary responsibility: helping construct a full-grown human being from a single minuscule egg cell, and then helping maintain us as we grow and traverse our life span.[26]

## DEEP SIMILARITIES IN THE GENETIC CODE

Biologists had long taught that the compound eyes of insects and the cameralike eyes of vertebrates are profoundly different in both structure and origin. It is true these very different eyes develop in very different ways, but geneticists have recently learned that the genes controlling the early development of the eyes of fruit flies are very similar to certain genes that control the development of eyes in mice. In fact, splicing mouse eye genes into an early fruit fly embryo causes the fly to develop extra eyes on its body. And the same thing happened when the eye gene from a squid was inserted into a fruit fly embryo. (The treated flies produced compound insect eyes, not mouse or squid eyes; these genes signal "make an eye here" and nothing more.) The ability of the genes to function in this way is strong evidence for a common origin hundreds of millions of years ago. One can imagine that an ancient common ancestor, perhaps a simple wormlike creature, had developed the genetic information for nothing more than bringing together a few light-sensitive nerves. Such "eye spots," and the genetic information that produced them, were the starting point for vision in a wide range of later animal lineages. The newly discovered genetic similarities in eye development suggest that the same original genetic information has been transformed and amplified independently through eons. This has produced a many-faceted image for insects and the single sharply focused image in the cameralike eye of the vertebrates and cephalopods (squids and their allies).

Earlier, geneticists had discovered that the genes controlling the early differentiation of the fruit fly embryo's body segments were very similar to some genes controlling early differentiation in the mouse embryo. Further studies have substantiated these disclosures regarding early development in such very different animals. Since fruit flies and laboratory mice have been studied for decades in genetics laboratories, it's not surprising that the newest revelations have come from our favorite laboratory subjects. All these recent findings support the commonsense notion that accidental duplications have been a major source for the development of new hereditary traits. With the original gene coding for a necessary part of the cell's machinery, the duplicated gene becomes available for potentially new and more elaborate biochemical tasks.

The study of DNA in genes and the protein sequences it encodes has strongly reaffirmed the view that all of life's rich diversity shares the same unitary historical origin. Every known form of life has its hereditary information transferred from generation to generation by DNA or RNA, using the same coding system to specify amino acids in protein construction. Thanks to this deep commonality, we are now able to "instruct" bacteria to produce human insulin or bovine hormones, by inserting human or bovine genes into their hereditary machinery. This facility came as something of a surprise. It had been thought that the genetic information would have changed profoundly over the more than 2,000 million years during which bacteria and nucleated organisms have been going their separate ways. Not so. Evolution seems to be fundamentally quite conservative. What has happened is that living things have retained similar basic functions, together with the genes that control them, over this entire time span. Mother Nature has apparently been following the old dictum: "If it works, don't fix it!" The fact that a human gene can tell the bacterial machinery to make an insulin molecule means our cells are still talking the same language and, perhaps more impressive yet, the "biochemical gears" still mesh perfectly.

Most proteins have portions of their structure that are central to their function. They also have parts that are not so important, with the capability of accumulating minor hereditary changes over time. For example, cytochromes in both bacterial and nucleated cells show considerable variation in some of their amino acids. One of these essential enzymes, cytochrome C, is made up of about one hundred amino acids. The cytochrome C enzymes of humans and rhesus monkeys differ in only one amino acid. Between humans and tuna fish the difference is twenty-one; between humans and wheat, forty-one. These differences in amino acid composition reflect mutations in the nonessential DNA instructions having accrued over time in the different lineages. Just as significantly, while humans, fish, and flowers share deep similarities in the structure of their genes, they also use the same intricate procedure to make more humans, fish, and flowers. We call this procedure sex.

## WHY SEX?

Organization of genetic information on the chromosomes was a central advance in the origin of life. In bacteria, after the single circular chromosome becomes duplicated, the two chromosomes become attached in close proximity on the cell membrane. The bacterium then divides precisely between the two attachment points so that each daughter cell receives one of the two duplicated chromosomes. Unlike higher organisms, bacteria have only a single set of genes on their solitary chromosome. Bacteria cannot carry recessive or dominant genes in the same individual as do plants and animals with two sets of chromosomes. In bacteria, asexual reproduction produces newly formed individuals identical to their progenitor, with two uncommon but important exceptions. The first is that of inevitable occasional mutations. The second exception involves transferring genetic information, both by exchanging chromosome fragments and by means of viruslike plasmids. These processes are as close as bacteria get to sex; but the benefits of new gene combinations are very real for the bacteria. It is only through genetic variation that living populations can propagate successfully in changing environments.

Things are more complicated in larger nucleated cells where the DNA strands are borne on larger chromosomes housed within the nucleus. In ordinary cell division (mitosis) of nucleated cells, all the chromosomes become two-stranded during the "quiescent" part of the cell's life cycle. As mentioned earlier, the cell divides so that, after division, each daughter cell receives a complete complement of one-stranded chromosomes. Things are a bit different in the process called meiosis, which forms the sex cells. Here the homologous chromosome pairs (one homologue comes from the female parent, the other from the male), each with two strands, align with each other in the center of the cell. By forming occasional crossovers between strands, the homologous chromosomes not only insure proper alignment before they migrate to opposing poles of the cell, they can also exchange strand segments. In this case, separation and division will give each sex cell one member of each *pair* of homologous chromosomes. Each daughter cell now has *half* the total number of chromosomes of an ordinary cell; a number that will be restored when male and female sex cells unite in fertilization. But why develop a distinctly different system (meiosis) to produce cells with half the number of chromosomes, cells that are useless unless they unite with another sex cell?

A central question in biology, one that still elicits considerable discussion, has been "why is sex so widespread in plants and animals?" Sex is a central component of the lives of all complex life-forms despite its obvious expense. Not only does a female have to go to all the trouble of finding a mate, but sex means that half the genes of her offspring are not her own, and half of her offspring will be males who may contribute little more than sperm to the next generation. Think of huge Alaskan brown bears where the males contribute nothing more than sperm to the next generation; the demanding responsibility of delivering and rearing the cubs falls entirely on the females. Wouldn't it be simpler just to let everybody reproduce themselves? In fact, some organisms, especially the smaller, simpler ones, do just that, but they tend to have episodes of sexual reproduction from time to time. Virtually all long-term studies of asexual self-reproduction indicate that eventually the lineage begins to decline in viability. Bad mutations accumulate, and there is no way of getting rid of them. Sexual reproduction and random mating, by constantly mixing up the chromosome sets and recombining genes, make it easy for natural selection to remove those individuals with deleterious mutations over time. (Natural selection is the process by which well-adapted organisms are more likely to contribute their genetic endowment to the next generation than poorly adapted organisms. This difference in reproductive success selects for those genetic characteristics that produce better adapted offspring over time.)

Just as important, early stages of meiosis allow for chromosome repair in a way mitosis does not. In addition, the pairing of homologous chromosomes in meiosis allows for "crossing over" where small sections of the chromosomes are exchanged. This provides the possibility for two good genes, carried on different but homologous chromosomes, becoming positioned on the same single chromosome. The individual who ends up with the chromosome having both "bad" genes is unlikely to produce successful offspring. (This is what natural selection is all about.) In effect, sex is both a repair system and a generator of continuing diversity. With the two chromosome sets (from the grandparents) being mixed up in thousands of different combinations in each parent's sex cells, there's little wonder why children of the same family can be so different. Asexual reproduction, in contrast, produces offspring who are all "holding the same hand of cards." Sexual reproduction produces a huge variety of different hands by continuously reshuffling the cards. Sex produces diversity within populations to meet the challenge of inevitable and unpredictable environmental changes, as well as having varied offspring who can take advantage of minor differences in the local environment. More significantly, the continual recombination of genes, thanks to sex, is perhaps the only way of dealing with the environment's nastiest challenge: short-lived, rapidly mutating pathogens and parasites. Finally, sex allows advantageous new genes to spread through the population much more rapidly than is possible in asexual reproduction.[27]

Sex has other interesting aspects. Exchanging genes with oneself or a close relative doesn't produce much in the way of variability among the offspring. Much better is exchanging genes with more distant members of the same species. Flowers provide many fine examples of ways in which plants have been selected to promote genetic diversity by having insects and other pollinators exchange pollen between distant plants. Sexual pairing in animals affords the opportunity for mate choice in many species. Fancy feathers vigorously displayed, elaborate rituals, or intense physical combat by males allows females to assess the health and fitness of potential mates. Female choice can be a powerful force in maintaining species fitness—what else might explain the peacock's tail?

Regardless of its many permutations, the universality of sex implies that it is profoundly significant. Long-term survival appears to be impossible unless a species' genetic heritage is kept in dynamic flux. Hounded by a horde of short-lived parasites and pathogens that can change their characteristics quickly, host organisms live in an especially dangerous and unpredictable world. By keeping the hereditary pot well stirred, sex has proven essential to survival in a nasty world.[28]

## ALLELES: DIFFERENT FORMS OF THE SAME GENE

The nucleated cell with two sets of chromosomes (one from Mom, one from Dad) has an additional advantage. With two complementary genes on the two homologous chromosomes, a "good gene" can do its work while masking the presence of its mutated or nonfunctional counterpart. These different forms of the same gene are called alleles. Thus, there is the possibility of numerous dominant and recessive alleles for the same gene; this is the basis of population variability. When we look at the great diversity among human beings around us we often ascribe it to "genes" but such phrasing, strictly speaking, is incorrect. The reason you and I are quite different is because we have *different alleles* of the genes all humans share.

One of the most dramatic examples of the effect of differing alleles is found in the mutation that causes sickle-cell anemia. Children receiving two sickling genes (one from Mom, one from Dad) develop the anemia and usually die before reaching adulthood. It is now known that the mutated gene differs from the normal gene in only one base-pair position (out of 861 positions on the gene), resulting in a single amino acid substitution on the hemoglobin molecule. This molecular change causes the red blood cells to develop a curved sicklelike shape, resulting in a reduced capacity to carry oxygen. But here the story develops an unusual twist. Some regions of Africa have populations with very high percentages of people carrying this dangerous mutation. Why didn't natural selection simply get rid of a gene that is so deadly in double dosage? The answer turns out to be in the *environment*.

These populations are found only in regions where virulent strains of malaria are endemic. In these areas people with ordinary hemoglobin (a good gene from Mom, a good gene from Dad) have a high likelihood of dying of malaria. Because of the effect on their red blood cells, it is the people with one good (normal) gene and one sickling gene who have the best chance of surviving *with* malaria. Natural selection had indeed been at work, selecting for those people who have both alleles—and thus maintaining high levels of the "bad" gene in the population. The bottom line is simple: the sickling mutant, which usually causes death in double dosage, increases survival in single dosage in environments of lethal malaria. Here is another example of Mother Nature's apparent indifference to our individual well-being. Sadly, losing up to as many as a quarter of its young people is the price these populations have paid, unknowingly, to survive in regions with virulent malaria. Africa is not the only area with such an unusual blood ailment; similar hemoglobin mutations occur in several other malaria-infested areas of the world. These examples show how advantageous it can be to carry two *different* alleles.

## LIFE: FROM OUTER SPACE— OR BY INTELLIGENT DESIGN?

For a very different picture of how life arose on our planet, claiming that bacteria and viruses have rained down upon the Earth from cometary clouds, check out *Our Place in the Cosmos* by astronomers Fred Hoyle and Chandra Wickramasinghe.[29] (Hoyle made major contributions to theories of stellar nucleosynthesis.) Theirs is a modern version of the ancient idea of "panspermia," in which the universe is perfused by bacteria and other forms of life. Unfortunately, these wildly imaginative authors don't seem to understand that most viruses have rogue DNA or RNA closely related to the DNA and RNA of many different specific animal and plant lineages. They even suggest that some human viral epidemics are of extraterrestrial origin. Worse yet is their suggestion that the "eggs" of metazoan animals could also be raining down on us from outer space. These authors don't spend much time explaining how viruses, bacteria, or the "eggs" could survive near-vacuum dehydration, subfrigid temperatures, and intense irradiation in interstellar space.

You might have been wondering why there has been no reference to viruses earlier in these discussions of the history of life. The reason is simple: viruses are not living things, or at least not *independent* living things. They cannot garner energy for themselves or reproduce themselves outside of the living cell of their hosts. Plant viruses and animal viruses are not at all related; plant viruses came from plant DNA or RNA, and animal viruses came from animal DNA or RNA. What most viruses are is bits of rogue genetic material that have managed to get themselves a protein coat to travel about, while having the "program" to target a particular kind of cell, enter that cell, and get that cell's molecular machinery to make more virus particles. They are, in a real sense, the world's smallest and simplest parasites. While some viruses may have very ancient roots, especially viruses that affect bacteria, most other viruses appear to be parasitic innovations. Viruses are a particularly nasty example of Murphy's Law. Here something went terribly wrong when, thanks to nothing more than a series of accidents, a short stretch of genetic information (DNA or RNA) became a deadly parasite.

The question of how life may have originated here on Earth is quite speculative, and we've touched on it only superficially. However, since we've considered panspermia, it's worth mentioning another interesting and scientifically unorthodox approach to this question taken by Michael Behe in his book *Darwin's Black Box*.[30] Behe, a biochemist, claims that the living cell is rather like a black box giving rise to cilia, flagella, and other very complex structures, which he sees as *irreducibly complex*. These irreducibly complex systems, he claims, cannot have evolved from simpler systems because they are utterly nonfunctional if any part of the system is removed. Like a mechanical mouse trap, dependent on specifically constructed parts that function only when brought together by their designer, Behe claims that such complex organelles or systems cannot have evolved from a more simple state because it is utterly unlikely that the different structural elements could have come together by chance at the same moment in time to produce so perfectly functioning a product. Such complex structures and processes, he concludes, like Rev. Paley before him, must be the product of intelligent design. Though such an interpretation will be deeply satisfying for those who believe in the reality of a creative deity, practitioners of orthodox science generally put currently "difficult to explain by evolution" problems on a back shelf and work on things that are easier to dissect and test. This highly pragmatic approach reflects a belief in eventually developing reasonable materialistic-scientific explanations. After all, haven't we explained many things that seemed miraculous and "irreducibly complex" only a few generations ago? Perhaps, with time, we will develop new insights and make new discoveries that will help us understand how such complex devices came into being, without calling upon supernatural forces.[31]

To recap the story thus far, the age of the earliest meteorites, when the solar system was young, is dated at about 4,560 million years ago. Life in its simplest form appears to have been operative here on Earth by 3,800 million years ago, around the time the early bombardment subsided. Stromatolites, probably formed by photosynthetic bacteria, made their appearance around 3,500 million years ago. The first fossil evidence for larger, perhaps nucleated, cells is dated close to 2,000 million years ago. After that, another 1,400 million years would pass before the fossil record suddenly displayed an exuberant diversification of animal life.

Present evidence suggests that the beginning of life on our planet occurred very quickly after the period of heavy bombardment. Of course, the earliest forms of life, however simple and inefficient, didn't have anything around to devour them. But then, more than 1,500 million years would pass before there is evidence for what may be the remains of larger nucleated cells. Such a long time period is consistent with the observation that the distinction between bacteria and organisms with nucleated cells represents the deepest dichotomy in the living world. The complex nucleated cell, powered by mitochondria, multiplying by mitosis, and having sex with meiosis, took more than a thousand million years to come into being.[32]

Surprisingly, the emergence of wriggly animals did not occur for *yet another* 1,500 million years. Perhaps these early periods of Earth's history, 3,000 to 600 million years ago, may not have been as boring as some have postulated. It's possible that our planet may have suffered repeated "snowball Earth" episodes. Under such conditions, only the very simple and the very small were likely to have survived.[33]

Overall, it took a bit more than 4,000 million years to transform what began as a hot, young, pelted planet into a world whose waters were teeming with actively swimming and crawling critters. Quite suddenly, between 550 and 530 million years ago, animal evolution shifted into high gear.

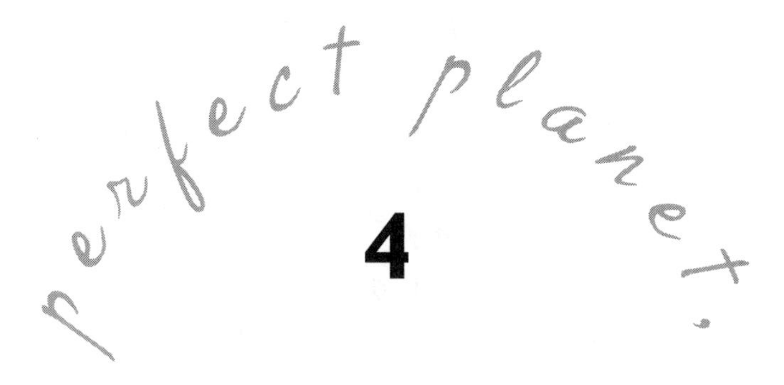


# 4

## LIFE EXPLODES
*The Last 560 Million Years*

clever species

Human uniqueness is the product of a long and fascinating history of life. To understand our uniqueness we need to appreciate the explosion of life-forms that populated the Earth long before humans came into being. Thus far, we've explored our star, our planet, and the first 4,000 million years of Earth's history—the time during which bacteria and complex cells arose. Interestingly, it was not until the period around 560 million years ago when the geological column first presents us with vague impressions of larger, more complex life-forms.

Difficult to interpret and unlike anything seen before or since, early larger life-forms lived in a fossil age called the Vendian (620 to 540 million years ago). The more striking creatures are referred to as Ediacaran fauna, after deposits in southern Australia (see Figure 1 on page 86). The majority of these fossils are somewhat flattened leaflike or radial pancakelike forms with smooth edges, some having a "quilted" construction with parallel transverse ridges—like little sleeping bags. Most are a few

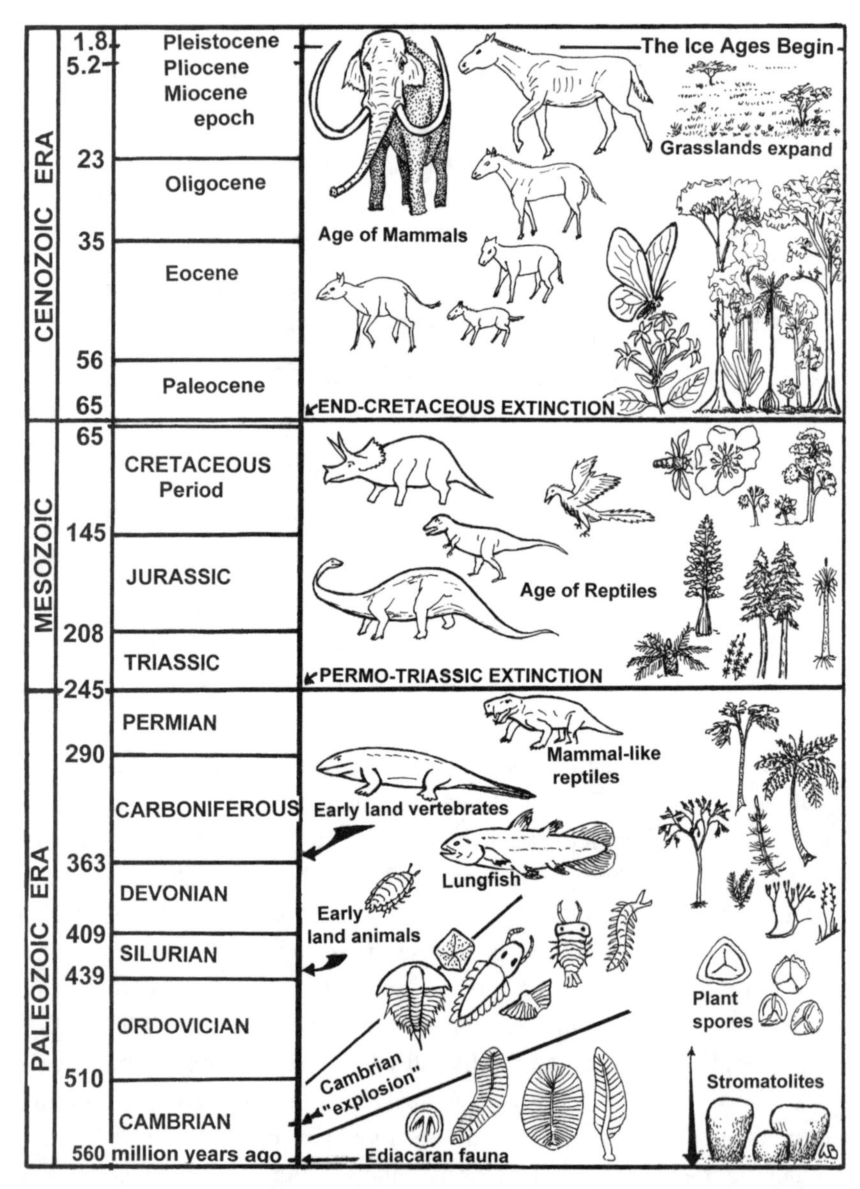


**Figure 1.** The last 560 million years of Earth's history. The Phanerozoic eon. Epochs are listed for the last 65 million years. Periods are listed for the time between 65 and 560 million years ago. Note that the time-scale for 0 to 65 million years ago is not the same as the time-scale for 65 to 560 million years ago.

inches long, but rare specimens grew to be a yard long. None is more than an inch thick. All lack external appendages; they had no mouths. It is possible these Precambrian (before 540 million years ago) creatures were floating jellyfish-like animals that had settled on the bottom shortly before being covered by sediments, to become part of the fossil record. These fossils come in a variety of outlines, but none of the creatures had a hard body, head, feet, spines, tail, or other distinctive features. They appear to have grown by adding segments at one end. More important, they appear to have been immobile, having left not a single trail. Also, there is no evidence of anything having taken a bite out of them.

An interesting possibility is that the Ediacaran biotas were restricted to cold, near-polar waters. Mikhail Fedonkin, a Russian paleontologist, points out that the rock matrix of these fossils resembles cold-water deposits, quite unlike the carbonate-rich rocks of tropical waters.[1] Many Vendian sites were inhabited by numerous individuals of only one or two species, just as is the case in some cold-water habitats of today. A cold-water habitat may also explain why Vendian fossils seem to show little bacterial degradation. Fedonkin also argues that the bacterial stromatolites dominating tropical waters at this ancient time may have created high levels of toxic pollutants. It seems reasonable to suggest that warm waters dominated by massive bacterial colonies 600 million years ago were inhospitable environments for early animals.

An oblong bilateral fossil occasionally showing a protruding undulating margin, *Kimberella*, has recently been interpreted to be a mollusk, vaguely resembling living chitons.[2] Again, there are no legs, no differentiated head, and no evidence of hard parts. After flourishing for perhaps ten million years, the Ediacaran fauna appear to have declined, superseded by a period of small "shelly" fossils and rare "worm trails." Until recently, these were the only good evidence for animal life from these early times.

Our understanding of early animal evolution changed dramatically in early 1998 when two reports of superbly preserved fossils were described from phosphoritic deposits in southern China.[3] This kind of fossilization preserves minute cellular details, but only of minute beings; none of these remains is more than a tenth of an inch in length. These fossils resemble modern red algae, minute sponges, and, get this, *embryo-like* forms similar to the embryos of some of today's living complex bilateral animals.

Dated at around 570 million years ago, these fossils represent a slice of time just before the Ediacaran radiation. (A sudden multiplication of living things is called a radiation.) An early appearance of sponges is consistent with their very simple organization, but there is no reason to assume that any of these early "embryos" actually grew into larger more complex animals. These new fossils have opened a door into a period that had previously revealed little of its mysteries. As Richard Fortey reminds us in his book *Life*: "The great narrative of geological time is a patchwork, a stitching together of odd fragments. . . . The past is continually erased, and the record of the most distant time survives only by a chain of minor miracles."[4]

## THE ORIGIN OF COMPLEX ANIMALS

Quite suddenly, between about 535 and 525 million years ago (a very short time period in our long history), the ocean world changed forever.[5] The rocks at the beginning of the Cambrian period, around 545 million years ago, began to show an increase in burrowing activities and trace fossils, followed by a mix of small, shelly fossils between 543 and 535 million years ago. Then came the "Cambrian explosion" itself. The first stage, at about 535 million years ago, includes the earliest mollusks, brachiopods (lamp shells), and echinoderms (starfish and sea urchins). By 530 million years ago the rocks also included arthropods (trilobites and crustaceans) and the ancestors of animals that would later give rise to scorpions, horseshoe crabs, and vertebrates. Within a period of less than ten million years, the rocks suddenly included the remains of many forms bearing heads, tails, legs, spines, eyes, and other useful features. Most of these new creatures left an indelible mark in the record of the rocks for a very simple reason: they had a hard, resistant covering. A crowd of hungry neighbors was the probable reason for developing a tough, fossil-friendly epidermis. Not only did they develop hard surfaces and defensive spines, some had sharp, rasping teeth. More significantly, the Cambrian explosion of animal fossils features a range of different morphologies. The period did not begin with generalized "protoanimals" that gradually diversified into different forms. Rather, the Cambrian explosion was marked by a variety of very different creatures from its beginning. Both coiled and clamlike mollusks, a variety of wormlike creatures, segmented trilobites, primitive crustaceans, chelicerates (ancestors of spiders, mites, and horseshoe crabs), starfishlike echinoderms, and weird forms having no modern counterparts, are all among this early cast of characters. As the Cambrian curtain rose, the stage was already populated by a variety of differing lineages.

Paleontologists have been puzzling over the mystery of this sudden and dramatic diversification for well over a hundred years. The Cambrian period had been preceded by extensive ice ages and major tectonic activity; but these events came long before the explosive origination of numerous new animal fossils. To recap: the larger, leaflike forms of Vendian time, the Ediacaran fauna, made their appearance about 560 million years ago, flourished for perhaps twenty million years, then declined. Next a "shelly fauna" of minute broken body parts is found in some deposits, quickly followed by the sudden explosive diversification of differing kinds of animal lineages between 535 and 530 million years ago. But whatever the actual progression, it seems unlikely that the sudden appearance of so many different forms at the beginning of the Cambrian period could have taken place without earlier and more fundamental biological innovations.

Today many diverse major lineages—phyla—of complex marine animals have minute larvae that develop directly from the fertilized egg. These floating larval stages are limited to less than a thousand cells in size. Their simple structure bears no resemblance to the mature form of the same animal. Within the minute body of these larval stages, a small number of undifferentiated cells are "set aside," and these are the cells that will give rise to the development of the more complex adult body plan. Though this type of two-stage development is widespread among diverse animals, a number of lineages have lost the larval stage from their life cycle. Because it is shared among widely different phyla of aquatic animals, the two-stage "indirect" development (with larval stage) appears to be the primitive condition, and the simpler "direct" development (without larval stage) is assumed to be a more modern or derivative condition. Human development is direct: the fertilized egg divides repeatedly to form a hollow ball of cells which then transforms itself into an elongate embryo. Whether fishlike, feathered, or furred, all animals with backbones (vertebrates) develop directly from the egg cell without passing through a larval state.

The two-stage sequence of indirect development has been called "a rather peculiar process," one that makes little rational sense—unless it was the original condition. It seems logical that the first animals, like modern microscopic rotifers, were unable to grow beyond a certain number of cell divisions. Setting aside a group of undifferentiated cells within the original free-floating larvalike animal initiated larger and more complex body plans, argue Eric Davidson, Kevin Peterson, and R. A. Cameron.[6] Their theory for the origin of more complex animals claims that changes in the regulatory processes underlying more complex development were possible only within these undifferentiated cells, after they had been "set aside" within these ancient microscopic animals.

The foundation of the second stage in the two-stage life cycle and the separation of different lineages of early animals may have taken place many millions of years before the beginning of the Cambrian explosion. Support for this view has come from recent analyses of molecular differences in living marine lineages. Based on assumptions of regular nucleotide substitution rates (so-called molecular clocks), these studies estimate that the major animal lineages may have diverged somewhere between 1,000 and 700 million years ago, long before the Cambrian explosion began. Such an origin would have afforded time for lineages to differentiate and to lay the groundwork for what would later become very different developmental trajectories, producing the basic body plans for animals ranging from jellyfish and clams to crustaceans and fish.[7]

As animals became larger and more complex, they required ever more sophisticated developmental protocols. To develop larger complex forms, cells must *know where they are and what they should be doing*, being informed by chemical messages from surrounding cells. Moreover, cells cannot proliferate uncontrolled; they must be tightly constrained within the developmental protocols that create a complex animal. Cancer is an example of uncontrolled cellular multiplication. Animal architecture is not created by magic; the information has got to be in the genes and put to use when and where needed. It is only in the last few decades that developmental biologists have become fully aware of how multifaceted a task this is.[8]

## BUILDING A COMPLEX ANIMAL

The evolution of more complex animals appears to have passed through a number of stages of increasing morphological complexity. The simplest animals in this hierarchy are the sponges that are multicellular and have more than one cell type, but do not really produce differentiated tissues. They do not have nerve cells, cannot propel themselves actively, and they have no plane of symmetry, though producing a large variety of three-dimensional forms. They are filter feeders. The second level of complexity is represented by jellyfish, sea anemones, and their allies, which are radially symmetrical, have an inner and outer layer of cells, and a saclike inner "gut" where they digest food. They do have nerve cells and many can swim actively. The third stage is best represented by creatures resembling flatworms (*Planaria* and others) that are bilaterally symmetrical, have bodies made up of three cell layers, and possess a complex nervous system but still have a gut with only a single opening. The next stage included the development of a well-defined internal cavity with a mouth at the front end and an anus at the other end, as exemplified by wormlike animals. Once these major features were in place (bilateral symmetry, three basic primary tissues, a nervous system, and an efficient flow-through intestinal tract) the stage was set for the orgin of a huge array of different kinds of complex critters, including velvet worms with leglike appendages, crayfish, centipedes, the ancestors of fish with backbones, and starfish (which lost their earlier bilateral symmetry). Hard skeletons were an important innovation, whether on the inside, like us, or on the outside, like lobsters. Inside or outside, a skeletal architecture could brace filtering or chewing systems, give leverage to muscles, make strong appendages possible, and produce a host of different animal forms.

Regardless of the complexity of the animal, it's important to remember that most larger plants and animals cannot make copies of themselves; they must revert to the single fertilized egg cell in order to recreate new individuals with the same complex mature form. But how do you program a *single* cell and its early daughter cells to create an entirely new organism? A complex animal requires a sophisticated series of regulatory genes to choreograph the transformation of a ball of cells into a three-dimensional creature with head, tail, legs, and other fancy accoutrements. Once "set-aside cells" had become a part of the organism, clusters of regulatory genes could begin to fashion larger and more complex body plans. Sponges, the simplest of animals, have at least one "*Hox* cluster" of regulatory genes. The slightly more complex, two-layered jellyfish and sea anemones have three sets of *Hox* genes. Annelid worms have eight, and humans have more than thirty. The deep similarities among these regulatory gene complexes between animals as different as jellyfish, worms, flies, and mice have been discovered only in the last twenty years, and they provide powerful evidence for the deep unity of all members of the animal kingdom.

Because the *Hox* (or homeobox) developmental genes are easily identified and very consistent in their organization, homeotic genes have been discovered and studied in a wide variety of animals. Knowledge of how these genes differ over many varied animal lineages has given us profound insights into early animal development. Among invertebrates, from worms to fruit flies, the homeotic genes are arranged in a linear sequence on a single chromosome. The strong similarities of all these linearly arrayed genes suggest that increase in number took place by gene-duplication events. In the fruit fly the eight homeotic genes form three series, corresponding to head, thorax, and abdomen, the three body regions characteristic of all insects. What the homeotic genes do is to control the timing, place, and degree of expression of other genes responsible for particular developmental mechanisms; and they do this in a temporal sequential order *following their order* on the chromosome. Here we have a basic key to the miracle of animal form: a sequence of developmental protocols. The fact that the simplest flatworms have few *Hox* genes, while insects and vertebrates have many more, is consistent with a more elaborate and complex body. Here's further evidence for "Murphy's Law" having played a major role in evolutionary advances, as gene doubling and other rare accidents allowed for the "creation" of more complex organisms built up by modular and reiterated parts.[9]

Vertebrates differ from insects in having four times as many homeotic or *Hox* clusters, each cluster on a different chromosome. Again, the genes are in a compact linear sequence on each chromosome, and this linear arrangement is sequentially concordant with unfolding development in time, as the early embryo differentiates. In all animals whose genetics is well studied, the *Hox* genes are in the same order as their expression along the body axis, beginning at the head and moving toward the tail.

These clearly orchestrated developmental programs transform an early oblong mass of cells into an elongate embryo with the characteristics of its lineage. Although early embryos may vary greatly (remember some have free-floating larval stages), all must go through a tightly organized series of developmental steps orchestrated by *Hox* genes that will produce the rigidly constrained body plan of the mature animal, whether crab, crow, or kangaroo. Later in development, further gene programs differentiate the fly from the butterfly, and the antelope from the anteater.[10] Now back to the Cambrian explosion.

If the early estimates for animal-lineage DNA diversification are correct, and if "set-aside cells" and the first *Hox* genes originated that long ago, we are still confronted with the question of why the Cambrian explosion took place so much later, and within a period that may have been as short as 5 million years. What caused all these different lineages to produce larger, better protected, and more complex forms at almost the same geological moment in Earth's history?

## WHAT IGNITED THE CAMBRIAN EXPLOSION?

The powder was there but what lit the fuse? Major worldwide environmental change seems to be the only reasonable scenario allowing so many fundamentally different animal lineages to proliferate at the same moment in time. Whether these creatures built their hard parts of carbonates, silicates, or phosphates, *all* diversified explosively during this same short time period. Not only are earlier complex hard parts absent from the fossil record, there are few trace fossils. There are very few trails of animals having dragged their bodies across the sand and no little footprints; a little later in the geological column both show up with fossils of more complex animals. But why all at once; and why all together?

The simplest and most plausible theory for the Cambrian explosion claims that life itself had changed the world. Beginning 3,000 million years earlier, blue-green bacteria had been busy splitting water by photosynthesis; later the red, brown, and green algae added their efforts to this activity. Using the hydrogen atoms of water to fix carbon and build energy-rich compounds, photosynthetic bacteria and algae released free oxygen into the sea and air. Over hundreds of millions of years this newly released oxygen would have been removed from the atmosphere by oxidation reactions across the Earth's surface and within its oceans. An unusual aspect of geological history is that nearly all the world's banded-iron formations were deposited between 3,700 and 1,800 million years ago, peaking between 2,300 and 1,800 million years ago. With the oceans slightly acidic, thanks to an atmosphere rich in carbon dioxide during these ancient times, a huge amount of ferrous iron was held in solution in the ocean. As free oxygen became available in the atmosphere and oceans, this ferrous iron reacted with oxygen to precipitate as ferric iron in a "grand rusting" of the planet. Today these ancient banded-iron formations represent about 90 percent of the world's iron reserves, and help sustain our modern technological society.

A distinctive geological deposit of these ancient times is pyrite-containing conglomerates, especially common between 2,800 and 2,100 million years ago. Their rounded shapes indicate these rocks had been bounced around in streams and rivers. If this were to happen today, the pyrite inclusions would react with oxygen in the aerated water and quickly be dissolved. The absence of pyritic conglomerates after 2,000 million years ago indicates that oxygen levels had risen to an amount sufficient to react with the pyrite by that time. Meanwhile, thanks to the busy little chlorophyll molecule, bacteria and algae kept right on belching forth more oxygen. It seems plausible that, after several thousand million years, most exposed oxidizable rocks would finally have become thoroughly "rusted." In addition, sulfur would have transformed into sulfates, and free carbon would have become oxidized as carbon dioxide or precipitated as carbonates. Once these processes played themselves out, oxygen levels in the atmosphere would have begun to rise slowly to higher levels. When oxygen approached a concentration close to today's 21 percent of the air, larger and more active forms of animal life became possible.

Once oxygen reached a high enough concentration, in the atmosphere and by solution in the oceans, animals could burn their dinners at a much faster pace. Now they could grow to larger dimensions and become more active. Filtering water and moving from place to place no longer required ocean currents. With more energy available, the little beasties could now pursue each other. The world was suddenly the host of a diverse array of worms, snails, clams, crustaceans, and other assorted creepy crawlers. Minute larvalike animal lineages that had differentiated tens of millions of years earlier were now free to produce larger and more complex life-forms. They did this explosively in an environment that was open and, at first, free of predators or competitors. The world would never again see such a rich and rapid diversification of animal life.[11]

The Cambrian explosion also included soft-bodied forms of strange and bewildering shapes. We first learned about these animals thanks to a wonderful fossil locality in Canada's Rocky Mountains, known as the Burgess Shale. Here, fine silt appears periodically to have cascaded down over shallow-water fauna, burying animals in an oxygen-poor matrix, preserving both hard- and soft-bodied forms. These deposits have revealed animals never seen before or since. Here was a time when Mother Nature was still in an early experimental stage, creating many bizarre life forms that have left no descendants, as well as early members of successful phyla that are still with us today. The variety of different basic animal body plans at this time appears to have been greater than those living in our oceans today. Perhaps some were "poorly designed" and could not survive the competition of more crowded seas, while others were simply the victims of bad luck. Dated at about 515 million years ago, the Burgess Shale gives us a picture of life's diversity in the latter half of the Cambrian period, about twenty million years after its explosive beginning.

The creatures of this extraordinary fauna, their discovery, and their importance for our understanding of the history of animal radiations are enthusiastically described by Stephen Jay Gould in his book *Wonderful Life*.[12] These same fossils are interpreted as representing a less diverse fauna and with a greater emphasis on convergence by Conway Morris in *The Crucible of Creation*.[13] Where Gould sees random chance extinguishing the rich diversity of the early Cambrian, Conway Morris focuses on adaptation as the more important factor, but he is unnecessarily harsh with Gould's argument. I suspect that the most reasonable interpretation of early Cambrian diversity will include elements of both positions. (Both authors summarized their views in the December 1998 issue of *Natural History* magazine.) Adding to all this excitement is a major new Cambrian fossil locality at Chengjiang in eastern Yunnan province, China. First excavated in 1984, these strata may match the Burgess Shale's, both in diversity of life-forms and in excellence of preservation, giving us an earlier view of Cambrian animal life around 525 million years ago. Both fossil localities make clear that the Cambrian was a unique period of animal proliferation. Its uniqueness is signified by a simple fact: in all the 500 million years of subsequent Earth history, only one major animal body plan or phylum, the bryozoans, has been added to those that originated in the Cambrian.

## MAJOR EXTINCTIONS: CHAPTER ENDINGS IN THE HISTORY OF LIFE

After almost two hundred years of intensive study by geologists around the world, the Cambrian explosion remains as one of the most clearly defined and abrupt events in the fossil record. The sudden and remarkable appearance of complex animals over a period of less than ten millon years is not an artifact of poor sampling or bad dating; it is a major worldwide historic event. Other abrupt changes have been recorded in the further history of life, as episodic events altered the faunas of the world. One might have thought that these abrupt changes in the fossil record, first discovered by geologists working mostly in Europe and the northern hemisphere, would slowly be "filled in" as careful paleontological studies added to our knowledge from other areas of the world. Filling in the gaps with new knowledge of transitional deposits and faunas has occurred in many sections of the stratigraphic column (and numerous gaps remain), but this has not erased or obscured the major chapter headings in the history of life. It is now clear that these great intervals ended in global extinctions, followed by a new entourage of creatures that signaled the worldwide initiation of a new fauna and a new period.

For more than a century, the earth sciences had been dominated by the idea that "the present is the key to the past." This paradigm, uniformitarianism, insisted that geological processes seen today offer a reliable picture of processes in times past; this view stood in bold contrast to the claims that a worldwide biblical flood was responsible for so many extinct animals and complex geological formations. Uniformitarianism has been the basis for major advances in the earth sciences and it continues to be an important principle. However, accumulating data suggest that our planet may, on rare occasions, have experienced sudden, calamitous events or periods of profound instability—unlike anything we see today.[14]

The geological period from the beginning of the Cambrian to the present is called the Phanerozoic, covering the last 560 million years. It is divided into three major eras: the Paleozoic (old animal life), Mesozoic (middle animal life), and Cenozoic (recent animal life); see Figure 1 on page 86. The Paleozoic began with the new Cambrian faunas and terminated at the end of the Permian period, about 245 million years ago. Just as the beginning of the Paleozoic is marked by a striking change from strata with few complex life-forms to a host of new animal fossils, its end is also defined by dramatic change: the Permo-Triassic extinction. (The Triassic period began the Mesozoic era.) Estimates suggest that as many as 90 percent of the species, half of the genera, and a third of the families of marine animals were wiped out during this transition. Among terrestrial vertebrates, 75 percent of the families became extinct. Huge outpourings of lava in Siberia occurred during *exactly* the same period in which so many animals became extinct; this widespread vulcanism may have had sudden and drastic effects on the atmosphere. It still isn't clear whether vulcanism was a principal cause or one of many factors in the Permo-Triassic extinction. Apparently, the event was not instantaneous and may have occurred in several pulses over several hundred thousand years. Although it is likely that a number of coincident factors may have been responsible, the Permo-Triassic extinction, 245 million years ago is clearly the largest extinction event in the history of complex life, and it's still a major mystery.

Just as the Mesozoic era began with the diversification of new faunas following the Permo-Triassic extinction, this period was terminated by another major extinction event 65 million years ago. Not as severe as the one terminating the Paleozoic, this particular extinction event is much better known because it is not as far distant in time and is marked by the end of the most spectacular group of animals to rule the Earth: the dinosaurs. For over 150 million years, dinosaurs and their reptilian relatives dominated the land, ruled the seas, and flew through the air. Except for a single side branch, the birds, no dinosaur survived the end-Cretaceous extinction. Although about 90 percent of the families of bony fish survived this extinction, *none* of the ammonites (relatives of squids with coiled shells) survived, ending their history of more than 300 million years. Many minute marine organisms perished at this same historical moment, and it took significant geological time before their numbers fully recovered. Among terrestrial vertebrates, 36 percent of the families went extinct, including *all* the dinosaurs, in contrast to the 75 percent of vertebrate families who went out of business in the Permo-Triassic extinction. More abrupt than the Permo-Triassic extinction, intensive work on every continent indicates that this extinction event took place at about the same time all over the globe.

The end-Cretaceous extinction is a fine example of how science deals with a puzzle. There were those who claimed dinosaurs died out gradually as climate changed toward the end of the Cretaceous period. The trouble with this contention is that climate is always changing, and, as you sample rocks closer to the extinction event, the probability of finding fossil strata over the shorter final time-span diminishes. Those who argued for a very quick and sudden extinction event had the further difficulty of correlating deposits from distant corners of the Earth and proving that the event had occurred at *precisely* the same moment in time. Fossils are often buried in loose deposits and are subject to reworking and redeposition over a wider stratigraphic range. Simultaneous extinction can be very difficult to demonstrate.

## THE ASTEROID (BOLIDE) HYPOTHESIS

The question of what had caused the end-Cretaceous extinction remained a back corner of evolutionary studies until a small group of scientists announced they had found evidence for a very dramatic scenario. In the 6 June 1980 issue of *Science*, Luis Alvarez and his associates claimed that an impact of extraterrestrial origin had resulted in wiping out the dinosaurs. You don't get much bolder than that in the halls of academic science. As evidence, the Alvarez team announced their discovery of an excess of iridium in the fine-grained clay layers found just above the time zone associated with the bones of the last dinosaurs. Their primary data came from sediments in northern Italy, as well as Denmark and New Zealand, where this particular time period was clearly recorded.

Beginning as an attempt to determine the length of time during which the grayish clay/shale layer marking the end of the Cretaceous period had formed, the Alvarez group assumed that the gradual accumulation of meteoritic dust might be a way to measure the length of this time period. What they discovered instead is now called "the iridium anomaly," a sudden increase in iridium concentration as one analyzed sediments across the end-Cretaceous horizon. Knowing that iridium is much more common in meteorites than it is in the Earth's crust, the Alvarez group claimed they had found evidence for an asteroidal impact.[15]

Science has been described as an activity making bold conjectures in the face of severe efforts to refute them. The asteroid-impact hypothesis initiated a frenzy of activity, both by those hoping to support the idea and those trying to demolish it. At first, papers for and against the hypothesis appeared in about equal numbers, but after a few years the balance began to shift. The iridium anomaly was found in what had been freshwater lakes, negating the counterargument that the iridium had come from undersea vulcanism exceptionally rich in iridium. Further work in Scandinavia, the western United States, and Australia made clear that the anomally was worldwide and had occurred at precisely the same time—within the range of error common and unavoidable to all geological dating methods. The discovery of shocked tektites (bits of glass and fused particles produced by high-energy impacts) was also associated with the iridium anomaly. They, too, are present around the world at this moment in time, with the largest tektites found around the Gulf of Mexico. In recent years, using deep core samples from oil-drilling operations, scientists are convinced they have identified the buried crater itself, the Chicxulub structure, along the Gulf of Mexico beneath the edge of Mexico's Yucatán peninsula.[16] As the evidence mounts, the bold conjecture appears to have been correct: the iridium anomaly is the signature of a catastrophic encounter with an asteroid.

Exactly how a meteorite between 0.5 and 6 miles (0.8–10 km) in diameter would affect the Earth on impact is the subject of wide-ranging speculation. Coming in at somewhere between 45,000 and 145,000 miles per hour, the impact would have caused an explosion of enormous energy. Smashing into the Yucatán peninsula's limestone rocks with the force of thousands of hydrogen bombs, it would have sent a charge of pulverized rock high into the atmosphere. This is the event that blanketed our planet with a layer of fine dust, minute tektites, and the iridium anomaly. At first there was some confusion about the exact locality of the impact, especially after a large concentration of the shocked tektites was found in Pacific Ocean sediments. Then scientists realized that, blasted high above the atmosphere, the tiny tektites fell west of the impact because the rotating Earth had spun eastward beneath them.

A globe-encircling cloud of fine dust would have had several harrowing effects. Scattering sunlight, the dust cloud would have darkened the surface of the earth, reducing photosynthesis everywhere. Today, it is represented in the geological column as a thin layer of grayish clay found around the globe. Some indication of the effect of high-flying dust can be inferred from the recent eruption of Mt. Pinatubo in the Philippines, which caused the average temperature in the Northern Hemisphere to drop by about 1°C (1.8°F) for most of a year. Though we will never know exactly how long the end-Cretaceous dust cloud persisted, it was sufficient to change the history of life. The asteroid hypothesis also helps us understand why some creatures survived and some didn't. The hypothesis implies that effects on animal life in the ocean would be especially severe. Minute phytoplankton near the ocean surface carry on photosynthesis, and if this source of energy fails or is diminished, life in the entire water column is severely stressed. Unlike forest or swamp, there are no thick layers of decaying vegetation at the bottom of the sea. Recent analyses of geological cores off the shore of New Jersey have found debris from the asteroidal impact above deposits containing the microscopic shells of many minute marine organisms. Above the debris layer nearly all the microscopic species are gone, topped by thousands of years of deposition before the fauna are fully restored. These cores, more than a thousand miles from the impact site, bear witness to a near-instantaneous death of marine planktonic creatures.

On land, most insect groups and small mammal lineages managed to survive the extinction event. But all the dinosaurs and their large energy-hungry relatives were wiped out. In contrast, some larger animals that can endure long time periods without food, such as crocodiles, turtles, and lizards, survived. Several distinctive bird lineages died out, but most got through to diversify into today's rich bird fauna. Land plants show some losses at the end of the Cretaceous period, but spore and pollen diversity remained about the same. Interestingly, some sediments exhibit a sudden increase of fern spores directly *after* the impact. This is not surprising since ferns are often the first plants to reestablish themselves after severe fires. All these observations are consistent with a short, devastating event. Today there seems to be little doubt: though dinosaur numbers may have been already declining in the face of climatic changes in the late Cretaceous period, their final exit was due to a catastrophic impact of extraterrestrial origin.[17]

The end-Cretaceous extinction event did more than wipe out the ruling dinosaurs; it reset the course of vertebrate evolution. Dinosaur dominion of the land had been complete; there were no other significant land herbivores or carnivores for more than 150 million years. After the great extinction episode, a group of small furry creatures, active mostly at night and kept in their place by the most fearsome carnivores of all time, was suddenly free of its oppressors. It doesn't seem likely that our lineage, the mammals, could have diversified significantly while dinosaurs were still abundant.

Here's another key historical accident in the history of life on Earth—and central to our own history. Shortly before the extinction, the fossil record indicates the presence of five modern orders of mammals (though there must have been more, as yet undiscovered). Ten million years later, twenty-five orders of mammals had enriched the diversity of our planet. Never again would the mammals proliferate with such exuberance. At about this time (early Eocene, 55 million years ago), North America had about 110 genera of mammals. It maintained that level of diversity until cooling trends diminished the fauna over the last 10 million years. Despite their origin more than 200 million years ago, it was not until after the terminal Cretaceous event that the mammals would come to dominate the world's terrestrial landscapes. By eliminating the ruling reptiles, the terminal Cretaceous extinction had opened the environment for mammals, allowing "our kind," to flourish and diversify. Most biologists are convinced: without the asteroid impact, we humans wouldn't be here.[18]

As our knowledge of geological history has improved, the reality and importance of extinction events has become more clearly evident, prompting an intense search for data in the literature that might give evidence for additional minor extinctions. That search led paleontologists Jack Sepkoski and David Raup to claim that they had found a regularity of about 26 million years between large and small extinctions. This dramatic revelation was quickly embellished by a few astronomers who postulated that the Sun might have a dark and invisible sister star, appropriately called Nemesis. This distant and unseen companion star would swing by the solar system every 26 million years, so the theory claimed, disturb cometary debris in the outer regions of the solar system, and send that debris raining down on us to produce the regularly scheduled extinctions. Fortunately, further calculations suggested that the 26-million-year periodicity was an artifact of the way in which the data had been analyzed. Nor was there any evidence for a dark Nemesis. This little episode reminds us that many scientific hypotheses are themselves subject to extinction (that's how science works).[19] But enough of extinctions. Let's get back to "creative" evolution.

## THE EVOLUTION OF A LAND FLORA

Animal life has been the primary focus of this chapter, but several important plant innovations during the Phanerozoic were major events in the history of life. Of these, the most significant was the initial development of a land flora. While ten out of thirty-five living animal phyla have been able to move from the sea onto the land, only a single green plant lineage was able to make this difficult transition, and it is a lineage that is utterly unlike any of its aquatic precursors. Incapable of moving to more favorable locations, land plants had to become adapted to desiccation, intense light, wind stress, and severe temperature changes in thin air. Though animals such as spiders, insects, and land vertebrates produced some unusual new forms on land, each represented the modification of a previously existing body plan. Land plants, in strong contrast, produced a series of entirely new and more complex body features as they adapted to the land and diversified.[20]

Plants and animals are very different beings. Except for photosynthesis, plants and animals share all the same basic life activities within their cells: respiring, dividing, and doing all those things characteristic of the nucleated cell. But plant and animal cells differ fundamentally in how they are packaged. Generally larger than animal cells, plant cells are bounded by rigid cellulose walls, constraining them to a stiff and static lifestyle. Plants may flex and move with the wind, but their individual cells cannot flex and move the way the cells of animals with thin membranous walls do. Thin flexible walls of animal cells enable them to change shape and even disassemble themselves when necessary—critical aspects of early animal development. Plant development is altogether different: growth and division are confined to special growing tissues, the meristems. Cells in the meristems—at the tips of roots and twigs—retain all their embryonic possibilities, continuing to produce both new cells like themselves and those cells that will develop specialized tissues. It makes no difference whether the plant is a month-old weed or a giant redwood thousands of years old; dividing cells at their growing points remain embryo-like in function. Because of their stiff, inflexible cell walls, plant architecture is fashioned from the subtle elaboration of these microscopic little bricks, firmly glued together and dividing only in a few special locations. Locked into a static architecture, most plants spend their entire lives firmly anchored in one spot, capturing sunlight to energize their activities. Though not very lively, they are the foundation of terrestrial ecosystems. Animals, meanwhile, must search out their energy sources, devour, and digest them.

Blue-green bacteria probably were widespread on damp soil and in shallow pools for hundreds of millions of years. Later, together with algae and fungi, they may have formed mats in moist surfaces. Because they did not possess resistant cell surfaces, these early forms of a land flora have left no trace in the fossil record. The first land plants were probably flat liverwort-like plants whose thin little leaves, like the land flora before them, left no record in the rocks. But they did produce decay-resistant spore tetrads (groups of four spores); these are the first indication of complex land plants in the fossil record. Such spores make their appearance about 470 million years ago and are the only indication of land plants for many millions of years. (Algae produce no such spores, and fungal spores are quite different.) It was not until about 410 million years ago, with spores diversifying, that the rocks preserved the first evidence of plant tissues. Among today's land plants, all produce lignified tissues *except* the liverworts, which may explain why the first 60 million years of land plant evolution left little trace of their form or structure. After incorporating strong, decay-resistant lignin and more efficient cellular plumbing systems, vascular plants soon created a richly three-dimensional land flora. Though the earliest fossil vascular plants were only one to six inches tall, diversification came quickly. By 375 million years ago swamp forests with one hundred-foot-tall trees had added a new dimension to the world of biological diversity. In less than 100 million years, plants had changed some of Earth's land surfaces as profoundly as the animals of the Cambrian diversification had changed the sea.

It seems likely that complex land plants could not have proliferated without the presence of soil fungi (fungi include the molds, mushrooms, and their kin). Fossilized roots of the earliest land plants show evidence of association with fungal tissues. Even in today's world, a great majority of land plants benefit from associations with soil fungi; the fungi make soil nutrients more accessible to the plant, while plants provide energy-rich compounds for their associates. Fungi were once thought of as plants themselves, but they are now placed in their own independent kingdom (coequal with the plant and animal kingdoms). They differ significantly from plants in that they lack chlorophyll, having walls made of chitin (not cellulose) and having a unique form of reproduction. They also differ from land plants in having a basically filamentous structure, lacking both embryo and meristems. They are essentially threadlike beings that occasionally produce larger structures for spore dispersal, such as mushrooms. Constructed of intertwined filaments, mushrooms, toadstools, and bracket fungi develop very differently from the way in which plants and animals are formed. Also, their nutrition is more like that of bacteria; they live *within* their food (except for their spore-forming bodies), exude digestive enzymes into the tissues around them, and then absorb the released nutrients through their thin walls. With the appearance of wood and lignin, the terrestrial environment could have run into a serious problem: an ever-growing pile of dead lumber. Fortunately, both bacteria and fungi can digest cellulose and lignin, the stuff that makes wood our favorite building material. Without fungi and bacteria, there would be no way to recycle fallen trees and plant litter back into simple organic molecules, available to nourish other forest dwellers.

In regard to land animals, the earliest fossils are tiny bits of cuticle, chemically extracted from rocks about 430 million years old. The first land creatures could have begun by feeding off sea weeds and dead sea creatures along ocean shores; later they could have ventured above the tidal zone to forage among algal mats and a mostly two-dimensional vegetation. As larger plants reached for the sky, their increased height and branching created a new three-dimensional environment providing protection from the Sun and wind, transpiring humidity to the air around them, and producing a rich layer of detritus. But expanding land floras did more than provide a rich new environment for land animals; they also appear to have changed the weather. By reducing erosion and sequestering carbon in coal swamps and underground roots, land plants drew down the concentration of carbon dioxide in the atmosphere. And with less carbon dioxide trapping heat through the greenhouse effect, the Earth's surface temperature began to cool, resulting in the Permo-Carboniferous glaciation. Lasting from 310 to 285 million years ago, this was the most extensive and longest glacial period recorded during the last 500 million years.[21]

Necessary innovations for becoming a land plant included a tough, waxy surface that would prevent evaporation of internal water through outer surfaces, erect strong tissues capable of dealing with wind and gravity, roots that would hold the plant in place while absorbing water and nutrients, and an internal plumbing system to distribute fluids. But waxy surfaces presented a serious problem: by preventing water loss, they also restricted the entry of carbon dioxide into the plant and the escape of oxygen. The problem was solved by minute surface pores (called "stomates") that could be opened to allow gas exchange through the waxy surfaces and closed during dry weather. An earlier and more fundamental innovation for plant life on land was the small disseminule (spore). This was capable of being carried long distances by the wind, was resistant to drying, and had the potential to grow into a new plant when conditions were favorable. The thin little plants might perish when the soil dried out, but with such spores they could reproduce and travel with the wind. Spores are something no marine plants have produced, and this confronts us with a fundamental question: why leave a comfy aquatic environment?

## WHY LIVE ON LAND?

Why might plants and animals have adapted to a severe land environment, when remaining in a comfortable moist liquid medium seems much the simpler strategy? Keep in mind that natural selection operates only in the here and now. Adaptations for some possible future environment are utterly unlikely or rare accidental side effects. To be selected, mutations must have a contemporaneous beneficial (or at least not detrimental) function. The most reasonable scenario for adaptation to life on land is that both plants and fish living in shallow inland waters were being confronted by short periods of dessication as their creeks, marshes, and mud puddles dried up. It is important to remember that many tropical regions have long dry periods each year, resulting in vegetation that, today, ranges from barren deserts to grasslands, thornbush, savannas, and seasonally deciduous woodlands. Even before we started chopping them down, rain forests made up less than 30 percent of the tropic's potential natural vegetation. All the rest is seasonally dry.

Tropical regions with limited rainy seasons have many rivers, swamps, and marshes that regularly dry up during the dry season. Such weather conditions create life-threatening challenges for the plants and animals that are adapted to these normally wet habitats. As their watery home dries up, modern East African lungfishes burrow deep into the mud, secreting mucus that forms a thin cocoonlike enclosure in which they can breathe air and survive for several years. Long ago, as their habitats dried up, our fishlike ancestors probably used their four strongly lobed fins to push themselves to a nearby stream. Based on a recent study of the fossil *Acanthostega*, the earliest amphibians had developed their skull and forward skeleton for greater movement of the head, better ventilation of their lungs, and better locomotion, though they still retained rather fishlike hindquarters. Perhaps their stronger front appendages developed as a response to selection for moving through thick underwater vegetation or crawling over sunken logs. But again, the probable reason for lifting one's body free of the support of water and moving across land was survival when stream channels or ponds dried up. Having become adapted to muddy, oxygen-poor water, the fishes ancestral to land vertebrates had already elaborated lungs for breathing air. These fish had four strongly lobed fins, which became transformed into four legs in early amphibians about 365 million years ago. (This explains why all us land vertebrates are four-limbed.) Most likely, this critical transition took place in fresh or brackish water environments where, unlike the ocean shore, dessication was a frequent threat.

Similarly, land plants are most closely related to a group of living algae found *only* in fresh water. For plants, adaptation to the land probably began by growing quickly during the wet periods in shallow pools and producing spores that could survive later dessication or be dispersed to other moist sites. The notion that the regularity of ocean tides and a rich tidal seashore environment was the springboard from which animals became adapted to the land may hold true for small invertebrates, but it seems unlikely in the case of plants and vertebrates. For life that had originated and evolved in the sea, the land was an extremely hostile environment.

## LAND PLANTS DIVERSIFY

Not only did the different lifestyles of plants and animals present them with differing problems in adapting to a land environment, their contrasting natures have also had far-reaching effects on their further evolutionary histories. The fossil record indicates that competitive interactions between plants are very different from such interactions among animals. Clearly, the most striking animal radiations have taken place into "open" predator-free space; first in the early Cambrian before predators were commonplace, and since then after both major and minor extinction events. Recent examples of exuberant speciation, such as fruit flies in the Hawaiian islands and cichlid fish in the East African Rift Valley lakes, have also occurred in geologically "new" habitats with few competitors.

Paleontologist Niles Eldredge has claimed that extinctions have been critical to initiating new biological radiations,[22] but the major diversification episodes of plants show no such effect. As far as plants are concerned, animal extinctions seem to have been irrelevant. Rather, major bursts of plant diversification followed significant adaptive innovations. Using those innovations as markers, we can divide land plant evolution into four major episodes. The first was the initial adaptation to the land environment by simple, prostrate, liverwort-like green plants, remembered in the fossil record only by their spores. The next major step required the development of stomates and efficient internal water conduction systems, allowing the growth of larger, erect body plans and woody vegetation. By 380 million years ago, favorable localities in warm, moist climates became the home of lush forests. The third step followed quickly: seed plants made their appearance about 360 million years ago. With the appearance of pollen and ovules, the sperm cells of seed plants no longer had to swim to distant eggs. Like their aquatic ancestors, liverworts, ferns, and mosses require thin films of liquid water for sperm to wriggle their way to the egg cell. Carried by the wind through the air or transported by animal agents, pollen grains emancipated seed plants from requiring water as a fertilization medium. Germinating near the ovule, pollen tubes can grow toward the egg cell and effect fertilization. With further growth, the ovule transforms itself into the seed.

Because seed plants no longer required a watery environment for sexual reproduction, their origin resembles that of amniote vertebrates (reptiles and their early sister groups). For these vertebrates, internal fertilization and eggs protected by a leathery shell eliminated the need to find ponds for reproduction and early development. As a boy, I once came upon an extraordinary scene: a number of pollywogs (immature frogs) were swimming about in a small puddle. What made this puddle so special was that some pollywogs looked like blunt-headed little fishes, even as their nearby, slightly larger cousins had already sprouted a pair of hind legs. Reptiles, birds, and mammals go through a similar early development, but they do it within the egg or mother's body. Fully emancipated from their aquatic ancestry, seed plants, reptiles, birds, and mammals were able to expand across the land.

The fourth and last major innovation in land plant evolution made its appearance in the fossil record about 120 million years ago. Though their mysterious origin may have been much earlier, a new and versatile group, the flowering plants, began its expansion and diversification early in the Cretaceous. Colorful flowers to attract insects and the enclosure of seeds within a special leaflike chamber are just two of the features that distinguish flowering plants. Expanding dramatically in importance on the world scene while the dinosaurs were still dominant, this new botanical lineage suffered little at the Cretaceous extinction event, and it has been diversifying explosively ever since.

As was the case in animals, gene doubling appears to have been a key factor in allowing plants to become structurally more complex over time. Increased numbers of homeobox genes are correlated with increased structural complexity in animals. Likewise, a larger number of MADS transformation factor genes is responsible for increased complexity in plants. Ferns produce nothing as fancy as flowers because they simply do not have as many MADS genes to create such elaborate structures.[23] But a really big surprise is that botanists' favorite experimental flowering plant, the little crucifer (*Arabidopsis thaliana*), has almost twice as many genes as that other laboratory standard, the fruit fly (*Drosophila melanogaster*).

Speaking of insects, we should note that they are the other super-successful land lineage. Their first major breakthrough was the development of wings for flight. Just how such a useful innovation may have gotten started is still a subject of conjecture. The second innovation, among the more modern insect groups, was dividing the life cycle into four distinctly separate phases. In these insects, the egg stage can be a very short phase or provide a long dormancy in which to pass a frigid winter or long dry season. The second phase, the larva, is devoted entirely to eating and achieving a mature weight. The third phase, the pupa, can also be a resting phase; it provides time for an extraordinary morphological transformation creating the final, winged adult, reproductive phase. Though complex, this sharply divided life cycle opened a Pandora's box of possibilities. Bees, butterflies, beetles, wasps, and flies all have this four-stage life cycle. It is those insect lineages with complex life cycles that have provided flowering plants with their most important pollinators. Together, flowers and their pollinators would create terrestrial ecosystems richer and more dynamic than any that had ever come before.

Our planet's land surfaces may have been populated by small, moist patches of bacterial slop for more than 1,000 million years, but until there was a rich three-dimensional land flora to support a feisty fauna, the likelihood of evolving smart monkeys and even smarter humans was zero. Creatures like ourselves are likely to evolve only when challenged by rich and complex environments. That's been true here on planet Earth, and it's likely to be true elsewhere in the universe as well.

We began this chapter discussing the explosion of complex animal life in the early Cambrian period. This period was followed by 60 million years (the Ordovician period) during which animal diversity continued to expand. Then diversity leveled off in the sea, only to be decimated by the Permo-Triassic extinction. Marine life slowly reached earlier levels of diversity, only to be thrown back once again by the end-Cretaceous extinction. And here again, the engines of diversification have been busy. Today's floras and faunas exceed the highest numbers ever reached before.[24] What accounts for these recurring patterns of diversification? Let's next take a closer look at the factors that have driven biological diversification to produce the bounty that we enjoy today.

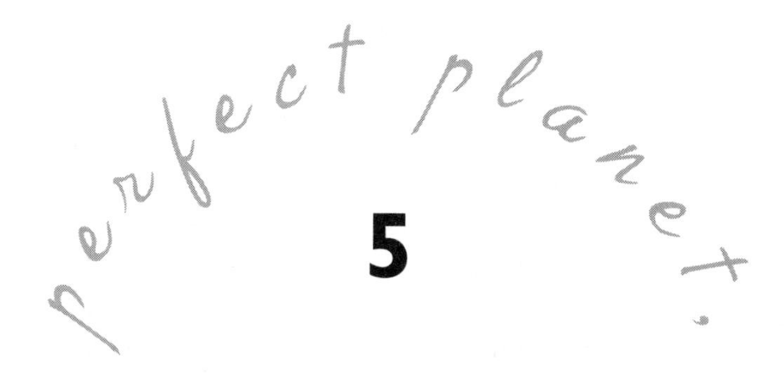


# 5

# WHY ARE THERE SO MANY KINDS OF PLANTS AND ANIMALS?

Creatures that can contemplate the heavens, construct theories, and build instruments to scan the deep cosmos are likely to evolve only on a planet rich in biological diversity. That's what happened here on Earth. Plant and animal diversity in the woodlands and savannas of Africa sustained the emergence of humankind. From among millions, a few animal species and a modest number of plant species have played central roles in the development of human societies in distant corners of the world. A rich terrestrial biota is a necessary condition for both the origin of intelligent life and its progressive development. And what a glorious diversity embellishes this fortunate planet, with plants and animals numbering in the millions of species. In fact, we are still discovering strange and wonderful new kinds of plants and animals.[1] Nevertheless, a few people respond to this extravagance by asking, "Why aren't there more?" Such a question may seem reasonable for those who haven't ventured far beyond their shopping malls, but in the real world long-term survival is the exception, not the rule.

Recently, twenty species of frogs and toads have disappeared from Costa Rica's Monteverde Cloud Forest Reserve, apparently the result of two unusually warm years, thanks to a strong El Niño episode. Many of these species, including the famous golden toad, were known only from this forest and are now presumed to be extinct. Such "ordinary" local extinctions, together with the geologically documented grand extinctions, make the survival and proliferation of so many living species in today's world a central question for biology. In fact, life in some areas is so rich and so varied it is difficult to understand how these regions manage to support literally thousands of coexisting species. That's the contemporary puzzle, focusing on proximal ecological relationships. Then there is the historical question: how did this glorious diversity of life come into being in the first place? Let's begin with the here-and-now questions of contemporary diversity.

Distant parts of the world (Australia vs. Africa) or different habitats in the same region (lowlands vs. mountain tops) support very disparate assemblages of species, and this seems easy enough to explain. Millions of years of separation on separate continents or islands, or consistently different temperature and rainfall patterns, can account for the different faunas and floras we find in different regions of the world or at the top and bottom of a mountain. But how is it that a small midwestern prairie can support as many as three hundred distinct plant species? Or, more mind-boggling, how can a few acres of tropical lowland rainforest support a similar number of tree species?

## TERRESTRIAL SPECIES RICHNESS

An important generalization is that most of the world's species of plants and animals live on land, despite the much larger area and huge volume of ocean water. Life in the sea has a far greater diversity of basic animal life-forms (thirty-five phyla) than life on land (ten phyla), but here we are speaking about the numbers of species of plants and animals. Some estimates suggest the overall species count may be ten times higher on the land than in the sea. Insect species alone probably outnumber *all* described species living in the sea.[2] The uniformity of aquatic ocean environments and their worldwide interconnections stand in strong contrast to terrestrial ecosystems. Over a third of the world's 18,800 species of fish live in freshwater rivers and lakes, despite the fact that only 1 percent of our planet's surface is fresh water. Isolation provided by rivers and lakes is undoubtedly a major factor, accounting for so many freshwater fishes in so small a volume of water.

A great range of physical factors, from dry to wet, hot to cold, and lowland to high-montane, contribute to making the land surface one of many more complex ecological parameters than can be found in the sea. In fact, many near-shore ocean communities do not seem to have many more species today than they did hundreds of millions of years ago; the species may be very different, but they've apparently filled most of the same parking spaces.[3] In contrast, the history of life on land suggests a very dissimilar trajectory, and it is the richness of life in terrestrial communities that will concern us.[4]

The problem of accounting for the extravagant numbers of diverse species on our planet and how so many creatures manage to share the same environments has elicited many explanations. It is clear that for species to flourish, environmental conditions must not be severe. Antarctica and central Arabia's empty quarter are thus notoriously poor in species for obvious reasons. Likewise, areas with long dry periods or long frigid winters do not support species-rich communities. Areas with generous rainfall, warm to moderate temperatures, and fertile soils in regions not having suffered major disruption in recent times are the places where species richness is the greatest. When such regions also have the varied physical and geographic parameters offered by dissected mountain ranges, the numbers of species become almost astronomical. Little Costa Rica, with high mountains and ample rainfall, supports twice as many species of mammals as does all North America north of Mexico. Ditto for ferns. That same republic, about half the area of Ohio, is home to more than one thousand species of orchids.[5]

Costa Rica demonstrates an obvious biological generalization called the "latitudinal species diversity gradient." Simply stated, as you leave the poles and move toward the tropics, species richness increases. Of course there are exceptions; in biology there are always lots of exceptions. Hot deserts often exhibit fewer species than the vegetation of their moister but more poleward borders. Likewise, certain groups of plants and animals are more numerous in the temperate zone than in the tropics.

These include salamanders, caddis flies, aphids, and conifer trees, to mention just a few. But these are unusual; most lineages have more species in the tropics than in temperate areas. Two reasons seem to stand out. The first is environmental severity. The utterly misnamed "temperate zone" features a perennial killer: really cold winters. It is only above about 10,000 feet elevation that you can expect to experience a severe frost in the tropics, and it lasts only a few hours per day—unless you are at an even higher elevation. A predictable climate within a narrow temperature range all year long allows far more "ecological specialists" to survive in the tropics than exist in the so-called temperate zone. A great many tropical species are "habitat specialists" whose narrow ecological tolerances are reflected in small or restricted geographic ranges, allowing more species to be packed into a given area. Whether plant or animal, if you live in a Midwestern prairie you have to be able to get through an occasional –20°F (–30°C) between December and February, and a possible 102°F (40°C) between June and August. Lacking such extremes, the tropics should experience much lower extinction rates—with increasing numbers over time. Also, a species that can survive severe Midwestern temperatures over its life span is more likely to range over thousands of miles. There are wide-ranging species that can tolerate a variety of habitats in the tropics, but they are a much smaller percentage of the overall species count.

Then there's another important factor in the tropics: energy. The Sun's rays traverse our insulating atmosphere more directly in the tropics, providing more energy to the vegetation. In short, the tropics are not only a larger pie with more energy, but the slices can also be cut a lot more finely. In addition, there's history. Ice Age climate oscillations have reduced species numbers nearer the poles much more severely than in the tropics. All told, a warm, predictable climate, over millions of years, has made the tropics a more congenial environment for species diversification.

South America is one of the world's richest regions in regard to tropical animal and plant species, and the most important factor may be the Andes mountains. The northern Andes are estimated to have 90 percent of the mosses, 80 percent of the liverworts, and 70 percent of the ferns found in northern South America.[6] The Amazon and Orinoco lowlands may be rich in species, but they lack the variety of habitats found in a high and deeply dissected mountain chain. Also, the species of the Amazonian lowlands tend to have much wider ranges than many of the montane species. Together with the rain forests of the Amazon and Orinoco lowlands and rich biotas in seasonally dry savanna woodlands, thornbush, and grasslands, the still-rising Andes provide South America with an enormous range of habitats.

While a complex interplay of rainfall patterns and topography may help us understand why South America harbors so many species, there is still the conundrum of having a great many plants and animals living together in the same small forest at the same time. Recently, a small (three-acre) forest census in Amazonian Ecuador has enumerated five hundred different species of woody plants with stems more than 1 cm (0.4 inch) thick. Many of these plants have the capacity to become trees, and finding three hundred species of tall trees in such forests is not unusual in tropical rain forests. But how do so many species of tall trees manage to live together in the same forest?

## DISRUPTION AND DIVERSITY

Visiting the tropical lowland rain forest can be a real disappointment. It is often not what people expect. A cathedral-like interior formed by huge old trees may be very difficult to find. Instead, the rain forest is usually something of a mess. As usual, Nature has produced a more complex reality. Robin Foster, tropical forest ecologist, has been surveying Manu National Park on the eastern flank of the Peruvian Andes. Because the Manu River is slowly shifting eastward as the Andes continue to rise, Robin can compare forests that are forming along the river's edge with those forests that formed in a similar setting hundreds of years ago. He finds that many forest giants begin as pioneer species on the river's exposed banks, reaching their full height within a hundred years. The forest growing up around them, no longer torn apart by flooding or the meandering river, reaches its full majesty between two hundred and four hundred years. As the forest ages, the early giants begin to fall, vines increase and the forest begins to look disheveled. By six hundred years, the forest no longer has the grandeur or stature it once had, but it is a forest richer in species than at any of its earlier stages.[7]

Explaining the diversity of trees in tropical rain forests has been the goal of many tropical biologists. Dan Janzen an entomologist and conservationist working in Costa Rica, suggested that seed predation played an important role. His research indicated that a small cadre of beetles and other herbivores was present under virtually every larger tree in the forest, waiting to pounce on the fruits and seeds as they fell from above. Many of these insects are specialists feeding on the fruits or seeds of only a few tree species. On the basis of his work with seed-eating beetles, Janzen proposed that the farther a seed fell from the mother tree, the more likely it was to survive and grow to adulthood. Fungal diseases under the mother tree will have exactly the same effect. These scenarios should result in a "hyperdispersed" distribution for trees in a tropical forest, with trees of the same species consistently distant from each other. As ecologists began to survey the trees in tropical forests, they found that some species were, indeed, more distant from each other than random dispersal would produce. But there were many trees growing in clusters of the same species as well as a number of species found only in very special settings, such as ridge tops or wet depressions. Ecological studies make it clear that there must be a variety of mechanisms supporting species richness in these forests. What was especially interesting was that some surveys found a few single-species clumps of the same age. Apparently, these trees had come in and grown up at exactly the same time, as if the products of a lucky accident. Perhaps a landslide had occurred just before these conspecific seeds had arrived, or a large tree had fallen, suddenly giving a set of conspecific saplings the light they needed to grow into the canopy.

Studies of tropical forests, tidal shores, and coral reefs have all contributed to a mounting consensus: small-scale disruptions are essential for maintaining high species richness. Predatory starfish can keep mussels from proliferating and dominating a portion of the tidal zone. While competitive interactions are essential in helping maintain species-rich communities, tropical trees indicate that there are other factors as well. Half of the more than two hundred species of trees living on Panama's Barro Colorado Island appear to be forest generalists.[8] They found their place, not by being especially adapted to a particular niche, but by having been in the right place at the right time. Thanks to luck, unpredictable small-scale disruptions, such as treefalls and storm-produced windthrows, are a major factor in maintaining species-rich forests. Such disruptions also help explain how the seed of a new immigrant managed to germinate and grow in a forest that already had 299 tree species; it's mostly up to chance. On the other hand, massive severe disruptions reduce species richness. Forests in southeastern Asia subject to strong typhoons have fewer tree species than those out of the range of these violent storms. A great many studies point to the same conclusion: *minor* disruptions play a major role in maintaining dynamic species-rich communities.[9]

## HOW FLOWERS CHANGED THE WORLD

In a textbook on paleobotany, Norman Hughes makes a series of estimates regarding the increasing numbers of species of vascular plants over the last 300 million years. Vascular plants include the flowering plants, conifers, cycads, ferns, and their relatives. Larger vascular plants give terrestrial vegetation its complex three-dimensional structure, and they are a land habitat's primary energy source. Hughes estimates that there were about 500 species of vascular plants at any one time in the Carboniferous period, 300 million years ago. By 150 million years ago, he estimates a world flora of 3,000 species. At the end of the Cretaceous, 65 million years ago, Hughes's estimate rises to 25,000 species. Today's total of vascular plants numbers around 300,000 species, of which about 250,000 are flowering plants. If Hughes's estimates are anywhere near correct, they imply that the flowering plants have been the primary factor in a hundredfold increase of higher plant species over the last 150 million years.

If most of today's tropical wet forests are highly dynamic, we can understand why flowering plants are dominant in these habitats: many grow fast and reproduce quickly. Their means of dissemination range from minute and feathery seeds dispersed by the wind to tasty and nutritious fruits transported by birds, bats, squirrels, and monkeys. Their colorful animal-pollinated flowers are even more essential. Animal pollination allows for gene transfer between isolated plants of the same species that are far apart, where wind pollination would be ineffective.

Conifers (including pines, firs, cedars, and their relatives) are not flowering plants; nearly all are pollinated by the wind. Unfortunately, wind pollination is not conducive to diversity. One hundred distinct species of wind-pollinated trees simply cannot share the same small forest. With that many species, the chance of receiving pollen from a member of your own species via the wind becomes hopelessly improbable. However, if you're looking for tall trees and lots of lumber, you can't beat the ancient conifer forests of North America's Pacific Northwest. In this corner of the world, mild winters and dry summers favor the evergreen cone-bearing (conifer) trees that don't bother much about growing sideways. No other forests in the world pack so many tree trunks into the same area. But don't let that fool you. These are not rich forests; they support fewer than two dozen large tree species. Also, these are not dynamic forests. Hurricane-force winds are not known to occur along this region of the Pacific coast, and with ample rainfall there are few severe fires. Here is an environment where thousand-year-old trees dominate a forest that supports fewer species than a Midwestern prairie. Wind pollination simply cannot sustain the same kind of species-rich forests that flowering plants have created—together with their animal pollinators. Using insects, birds, and bats, which can seek out conspecific trees, the flowering plants can pack three hundred species of large trees into the same small area of tropical forest. Those three hundred different species, in turn, support a mob of herbivores, which have their own predators and parasites. Building complex communities is one way in which flowering plants and their animal pollinators have changed the world.

Flowering plants excel in other ways. Their wide array of growth forms range from moss-sized orchids growing on tree limbs, bushy shrubs, and giant banana-like herbs on the forest floor, to woody vines and trees reaching high into the canopy. Flowering plants play a similarly important role in seasonally dry tropical woodlands and savannas, as well as in our northern deciduous forests.[10] Flowering plants support not only a more diverse vegetation but also one that is far more nutritious. For example, two species of English oaks have been recorded as supporting 284 insect species for all or part of the insects' life cycle.[11] No conifer, fern, or cycad comes close to being such a generous host. Estimates of the numbers of insects supported by flowering trees in rain forests are even higher. By exploding insecticide bombs in the crowns of large rain forest trees, researchers have counted as many as 163 species of beetles and 43 species of ants on a single tree. Young leaves, sap, nectar, seeds, and, especially, succulent fruit are the major food sources for animals living high in the canopy. Fleshy fruits undergo a carefully cadenced development before they reach maturity. Starting out small, green, and bitter, they are transformed as their seeds become ready for dispersal. Quickly they become brightly colored, sweet, and succulent, ready to be eagerly devoured by potential dispersers.

Larger flowering trees in moist evergreen tropical forests are often festooned with many smaller mosslike to shrubby plants growing on their trunks and outstretched branches; these plants are called air plants or epiphytes. Perched high up where there is more light, most epiphytes can survive only where rain or misting clouds are frequent, as in rain forests and montane cloud forests. Many of these epiphytes are flowering plants themselves; here is where most of Costa Rica's thousand orchid species live. Epiphytes are not parasites; they simply sit on the trunks or high branches (unlike mistletoes or other parasites whose roots enter the host plant and extract nourishment). Forming "aerial gardens," epiphytes help support a fauna of small animals very different from that on the dark forest floor, adding greatly to the numbers of animal species living in the forest. As Edward O. Wilson explains: epiphytes are "diversity multipliers."[12]

Flowering plants also sustain diversity in regions with long dry seasons. Whether acacia thorn scrub in Africa, eucalyptus woodland in Australia, shrubby cerrado formations in Brazil, or the chaparral of California, tropical and subtropical deciduous forests and woodlands are often made up almost entirely of flowering plants. Here, the amount of annual rainfall may be the primary determinant of species richness. For example, there may be as many as eight species of lemurs in a single Madagascan forest. This number of coexisting lemur species is directly related to the diversity of trees in the forest, which in turn is largely determined by rainfall.[13] Inevitably, as rainfall declines trees become smaller and more widely spaced, and the vegetation is reduced to thorny thickets. And where dry seasons dominate most of the year over a flat or gently undulating landscape, a special group of flowering plants abounds—the grasses.

After a long, hot dry season, broad open grasslands can turn into infernos as fires race across the landscape. Exactly what grew on such sites before the development of grasslands is not clear, but it seems unlikely that such habitats could have been as full of life as a modern grass-dominated prairie, savanna, or steppe. Like most herbaceous plants, grasses are almost never preserved in the fossil record; however, the hard teeth of animals that have grazed them are frequently found as fossils. The diversification of grasses with hard silica in their tissues produced strong selective pressures for the evolution of longer lasting and deeply rooted teeth among grazing herbivores. The fossil record documents this change among fossil grazing mammals between 30 and 20 million years ago—thanks to the expansion of grasslands during that time.

With undeveloped growing points protected close to the ground, grasses can quickly sprout new leaves after they have been heavily grazed, or, more important, after dry-season fires have scorched the landscape. After fires, and once the rains begin again, the quick growth of grasses gives them a head start. Dominating these broad, flat, seasonally dry habitats, grasses are the foundation for large herbivore communities throughout the world. Interestingly, recent research in grasslands is showing that moderately but continuously grazed grasslands are richer in species and more productive than nearby grasslands where grazing animals have been excluded.[14] It was the expansion of savanna grasslands and their fauna in eastern Africa that played a central role in the development of our own lineage (as we'll see in chapter 7). From a contemporary perspective, it is obvious that flowering plants have helped make our planet the abundant zoo it is. But are there other characteristics of flowering plants that contributed to this bonanza?

Flowering plants are unusual in a number of ways. Not only do they include far more species than all other land plants put together, they also have a greater variety of growth forms than is found in any other plant lineage. Their life-forms range from giant barrel-trunked baobab trees to the tiny disklike plants of pond scum (*Lemna* and its allies). They may be unbranched like a coconut palm, or a maze of branches as in a creosote bush or a tumbleweed. Leaves can vary in shape from linear and grasslike to the size and shape of umbrellas. Lifestyles range from small desert ephemerals, completing their life cycle in a few weeks, to long-lived forest giants. This confronts us with the question: why are flowering plants so much more diverse in life-forms than other plants?

One answer to this question is that flowering plants have developed an unusual twist to the business of sex. In most organisms, sexual reproduction entails the union of sperm and egg cells, but flowering plants have added something. Here a single pollen grain produces *two* pollen nuclei: one nucleus fuses with the egg cell to form the new embryo in the manner of a sperm cell while the other nucleus fuses with two maternal nuclei to begin the formation of nutritive tissue for the growth of the embryo. This unique process is called double fertilization and has a special advantage. Double fertilization results in the production of energy-rich seeds *only after* fertilization has occurred. In most nonflowering seed plants, nutritive tissue is formed in the seed *before* fertilization; this is a huge waste of energy if fertilization fails to take place. There's no such waste with double fertilization. Here's the primary reason why no other seed plant lineage includes species of small herbs: little plants simply don't have the energy resources to build seeds that may never sprout.

In addition to a great variety of growth forms, there's also chemistry—flowering plants produce a huge variety of chemical substances. Our own species has made use of this cornucopia in many ways. The primary sources of medicinal, herbal, and flavoring agents for all indigenous societies are flowering plants. It seems likely that most of these compounds play a role in making the plant less palatable for herbivores. In turn, this has resulted in a diversification of insect lineages, with particular insects developing detoxification mechanisms for the defensive compounds of a particular plant lineage. Our common milkweed (*Asclepias syriaca*) is fed on by characteristic species of beetles, bugs, and caterpillars. These are species that you rarely see on other plants; they have the special ability to sequester the milkweed's poisons and they live almost exclusively on milkweeds. By sequestering the poison, these insects become poisonous themselves, their bright colors a warning for those who might be interested in eating them. This is just one example of how the chemical diversity of plants has promoted insect diversification through feeding specialization.

Of all the keen attributes of flowering plants, colorful flowers pollinated by animal vectors may have been the single most important innovation, and not just in building species-rich communities. The central challenge for any species or population, plant or animal, is "not becoming extinct." Obviously, populations with few individuals are more likely to go extinct than populations with many individuals. Small populations may suffer random gene loss and a diminishing of their genetic resources unless they can get new genetic contributions from more distant members of their species. And that's why animal pollination is so critical. Animals can traverse considerable distances in their search for nutrient energy. Once a bumblebee has found a good nectar source and figured out how to obtain the nectar from within the flower, it makes economic sense for it to search for other similar flowers, rather than to spend valuable time trying to figure out how to get nectar from a dissimilar flower. Thus "pollinator fidelity," where the insect visits flowers of only one or a few species, is nothing more than energy conservation for the insect.[15] By bringing pollen to small isolated populations, insects and other animal vectors can spell the difference between survival and local extinction.

Finally, flowering plants are significant for another very simple reason: most are substantial in size. Beetles may outnumber flowering plants, but you can easily cram a thousand specimens of most beetle species into a gallon jug and have space left over. For the big picture, all these beetles don't add up to a whole lot of biomass.[16] More fundamentally, nearly all flowering plants are *primary producers*, capturing the energy of sunlight to keep the food chain churning. By converting a good percentage of this solar energy into flowers, nectar, fruits, and seeds, the flowering plants provide a richer variety of food resources than ferns and other seed plants. Thanks to flowering plants, there has been more food for the table, and many more to devour it.

## SPECIATION, THE ENGINE OF DIVERSITY

Now that we've got a better idea of how so many plants and the army of animals they support manage to live together, we can tackle the more fundamental historical question: where did all these species come from? What might have divided an ancestral species into two different daughter species? The fossil record gives evidence for species that have changed slowly with time and developed new features but without splitting. This is called "anagenesis" or "phyletic speciation," and the differing (earlier and later) entities are sometimes called "chronospecies." In contrast, contemporaneous speciation results when a lineage splits to form two new independent branches or "clades." Such speciation produces additional new species. This is the "engine" that has allowed the world's biota to rebound from extinction events, both large and small. But before we discuss speciation, we need to have a general idea of what it is we mean when using the word *species*.

Species are defined by most biologists as populations or systems of populations that are genetically isolated from each other. Ernst Mayr's phrasing has been adopted by most zoologists: "Species are groups of actually or potentially interbreeding populations which are reproductively isolated from other such groups."[17] The word *potentially* gives us botanists a tough time since many species that give no evidence of exchanging genes in nature can be induced to interbreed and hybridize in greenhouses. Most pragmatic biologists avoid the "potential" part and base their judgments on what seems to be going on in the woods and fields.

Given enough time, lack of gene exchange should make two separated populations quite distinct in appearance. Fixation of various new mutations, random loss of differing genes, and slightly different selection pressures should all play a role in contributing to divergence. If separated populations interbreed after renewing contact, their differences will be bridged by hybrids, and they may reunite as a single more variable species. Mayr's biological (or genetic) species concept seeks to avoid this problem by defining species as those entities that will remain distinct over evolutionary time, forever unable to exchange genetic information, and that's the reason for including the word *potential* in the definition. The horse and donkey are good examples of species. Their hybrid, the very useful mule, is strong but sterile; and thus there is no way for genes to get from the donkey population into the horse population, or vice versa.

Unfortunately, there are serious problems with a species concept defined by genetic isolation. First of all, we have little or no information on actual gene flow in most wild populations, and so we have to work with suites of morphological or molecular distinctions that might indicate the presence or absence of gene exchange. Second, many plants and some animals have species that seem to be clearly distinct but give evidence of hybridizing from time to time. These can give biologists some real headaches, though it doesn't seem to bother the misbehaving species at all; so-called oak species in our own woodlands are an annoying example. Nevertheless, the idea that "good species" do not exchange genes with other species is very useful; it defines population systems that are on independent evolutionary trajectories. Such a definition also implies that the process of speciation must establish genetic isolation. Thus, the central problem of speciation becomes, what factors promote genetic isolation?

Darwin wondered how "the lesser differences between varieties became augmented into the greater difference between species."[18] He believed that the more divergent forms of a species would be more likely to succeed and better enabled to adapt to a niche that was beyond the reach of its parent species. But Darwin never really pursued the question of which factors were actually involved in the origination of new species. Today, there seem to be a number of reasonable answers to this question—answers that are by no means mutually exclusive. In some cases, animals appear to have diverged in their mating behaviors and thereby created new species no longer capable of interbreeding. Plants can sometimes double their chromosome numbers and begin new lineages in that way. However, these are special circumstances; a more general idea, favored by most biologists, is the notion of geographic isolation as being the most common cause of speciation.

Populations that have become isolated on islands, mountaintops, or by other means may diverge from their closest kin. The American haw fly, whose larvae feed on the small, applelike fruits of native eastern hawthorns, apparently gave rise to what is essentially a new species. This occurred when a few members of the original species began living on apple trees. Introduced from Eurasia, apples develop and mature about a month earlier than hawthorn fruits, and since the mature insects court and mate *only on the fruits* of their host plants, the new pioneers became isolated in time and space by having shifted to apple trees. Today the haw fly and its new apple-dependent sister species live as effectively isolated populations, an incipient species having developed in less than one hundred years.[19] This example brings to mind the two species of lice that have made humans their home. One species lives in the hair of our heads, the other in a more sensitive region. Could this have been the product of a similar speciation event?

An interesting demonstration of the importance of geographic separation for speciation on a larger scale is provided by ferns living on islands off the coast of South America. A careful analysis of endemic island fern species (those found nowhere else) indicated that they were related to *rare* continental species.[20] Apparently, when common species of ferns reached the islands, spores of that same species from the mainland would continue to arrive and establish new plants with a sufficient frequency to "keep in touch" with the mainland species. In the case where a rare fern species had established an island population, the likelihood of continued gene flow from the mainland was minimal and isolation was effective. Thus, it was only these species, rare on the mainland, that were likely to develop into new and endemic species on the islands.

Conventional wisdom insists that without interbreeding and given enough time, the geographically isolated population should evolve into a new species. Later, recontacting the original population, the new species may be able to interbreed with the species from which it arose. But the hybrids should prove sterile or dysfunctional, with the two species unable to exchange genes through their hybrids, as in the case of mules. This has been the classic scenario for geographic or allopatric ("other homelands") speciation, and there is no question that it has played an important role in producing much of the world's rich diversity. Unfortunately, the world has so many species of insects, mites, ferns, and everything else that geographic isolation seems an insufficient explanation.

A serious problem with the geographic isolation scenario is that some populations that have been separated for millions of years are still perfectly interfertile. The English oak (*Quercus robur*) planted on the streets of Boston has produced fertile hybrids with nearby American white oaks (*Quercus alba*), even though these oak species have been separated for more than 20 million years. Likewise, the Eurasian plane tree (*Platanus orientalis*) and the North American sycamore (*Platanus occidentalis*) are the parents of a vigorous and fully fertile hybrid widely used in city plantings. Longtime separation has done nothing to induce genetic isolation between these species. Similarly, lizards of Australia's northeastern rain forests, separated for millions of years, show no evidence of divergence or speciation over that time period.[21] The physical separation of species may result in genetic divergence over time, but there is no a priori reason why that genetic divergence should result in the loss of interfertility. Geographic isolation may be an important factor in speciation, but it doesn't begin to tell the whole story.

## ECOLOGICAL DIFFERENTIATION

In Iceland's largest lake, a small northern fish species, the Arctic char, has separated into four rather dissimilar forms: two, a larger and a smaller, spend most of their time at the lake bottom where they pursue snails. Dark in color, they are camouflaged for their habitat. The other two forms live in open water, one feeding on smaller fish and the other on crustaceans and insects. Careful genetic analysis indicates that all four forms came from a single introduction. Because this lake was frozen solid during glacial times, the differentiation of the four forms must have occurred within the last 12,000 years. Though they may not be fully differentiated species, they seem to be well along the road in that direction, without ever having been geographically isolated.[22]

More spectacular has been the evolution of an estimated three hundred species of cichlid fishes in central Africa's Lake Victoria. Because their jaws are articulated in an unusual way, with a second set of grinding teeth in their throats, these fish have been able to evolve a wide variety of distinctive lifestyles. Their names indicate what some of them do: fish eaters, snail crushers, algae scrapers, prawn eaters, eye pluckers, tail biters, fish-scale scrapers, and leaf choppers are among the list. Female cichlids hold developing eggs in their mouths, thus protecting the most vulnerable stage of their young's development. Sad to say, a few cichlid species have adapted to this situation and have been labeled the "kiss of death fish." These feeding specialists collide with the females in such a way as to suck the eggs from their mouths.[23]

A generalized body plan, care of the young, and monogamous pairing in many species have been other factors favoring cichlid diversification. The explosion of cichlid species in Lake Victoria took place in a lake which may be less than 200,000 years old—all this in a body of water without major geographical barriers. Here, too, the DNA evidence suggests an origin from a single introduction. Similar cichlid diversification with convergent trends in local speciation has taken place in Lake Tanganyika (about three hundred species) and Lake Malawi (over five hundred species). In all these east African lakes, ecological specialization in order to reduce competition seems to be the most reasonable explanation for such a profusion of recent speciation.[24] Sadly, human activities in and around Lake Victoria have resulted in more muddy, turbid waters, and the fish don't seem to be able to see each other as well as they used to. The wrong males and females are getting together and producing hybrids. Species that were separated and distinct are fusing into hybrid swarms, with a consequent loss of "species" diversity. Apparently many of these fishes had not fulfilled a central requirement of Mayr's biological species definition: they are still *potentially* interfertile.[25]

Careful study over three decades on Darwin's finches of the Galápagos Islands has clearly documented evolution in action. These bird species, distinguished largely by beak structure, life style, and food sizes, have demonstrated subtle changes in bill form and size as their populations fluctuate in size and survivorship over many years of drought, punctuated by the intense and unpredictable rains of El Niño events. In addition to changing in response to the rains and vegetation around them, the birds clearly reflect the influence of competition. Their beaks are similar when a species is the only seed-eating finch on an island, but the beaks diverge when two species inhabit the same island. Called *character displacement*, this is direct evidence for the role of competition in driving species apart. Also, competitive "pressure" for morphological change may come from very dissimilar members of the community. One species of Galápagos finch living in the northern islands is smaller and uses nectar as a more significant portion of its energy sources than do birds of the same species living on southerly islands that share those islands with bees; these bees are not present on the northern islands. Competition with bees for nectar appears to be the primary explanation for the differences in form and behavior of the southerly populations.[26]

During my first encounters with Costa Rica's flora, I worked on the genus *Piper* (which includes black pepper from South Asia). American *Piper* had a bad reputation because of the difficulties in distinguishing the species. Sure enough, when working with museum collections, there were many arrays of specimens I could not disentangle. But when collecting in Costa Rica's forests and roadsides, it seemed as if any child might tell them apart. What was going on here? Slowly, it became clear; the species most similar to each other did not grow together in the same habitat, though they might be found less than a kilometer apart. Once I knew their habitats, it became much easier to recognize Costa Rica's ninety-five species of *Piper*.

Working with other families of flowering plants in Costa Rica, I found many other examples of closely related species clearly separated by habitat or microgeography. In the most striking instances, two similar "sister species" grow in adjacent wet forests on the same mountain slopes but not at the same altitudes. One species grows only at cooler, higher elevation, while its relative is restricted to lower, warmer elevations—with no hybrids and no zone of overlap between them.[27] Birds often show a similar pattern, with narrow endemics confined to highland cloud forests and their closest relations in the warm lowlands. In nearly all these cases, the lowland species are wide-ranging, and the higher montane species are found nowhere else. It seems as if the montane endemics had "budded off" from their more wide-ranging, lower-elevation congeners. Because these plant and bird species are each others' closest relatives, it seems safe to conclude that they too are the products of recent ecological divergence.

The most important point about the ecological differentiation scenario is the suggestion that there may be strong selection *against* gene exchange in special situations. If a particular population is becoming specialized for a particular environment, genes from other populations adapting to dissimilar environments will probably be deleterious to the success of offspring. Hybrids between the differing populations may be poorly adapted to either the new or the old habitat. In this situation, anything that opposes interbreeding will be strongly selected for. Alfred Russell Wallace, cofounder of the theory of natural selection, suggested that species could be formed by selection in exactly this way.[28] The importance of his proposal is that it implicates universally present selective forces as a primary engine of speciation, in contrast to unlikely random events, such as broad-scale geographic separation. Keep in mind that the English and American oaks, separated for more than twenty million years, are still perfectly interfertile. Isolation alone cannot guarantee intersterility. Wallace's suggestion answers the question of how species originate in a way that Darwin and most recent biologists have not. Also, Wallace's idea addresses the question: why should any genus have genetically isolated species? With a discontinuous environment to which different populations must adapt, it makes sense that genetic isolation insulates each population from genes adapted to foreign and inappropriate conditions. From this point of view, the answer to the question "Why are there so many kinds of plants and animals?" is that the increase in numbers has been driven in large part by selection for ecological specialization. If a species is adapting to a cool montane cloud forest, genes from other environments are not likely to be helpful.

## MAKING SENSE OF DIVERSITY

According to the biblical book of Genesis, one of the first things Adam did was to give animals their names. Just as actions and concepts require words to describe them, the important elements of our environment have to be named and characterized so we can convey information about our surroundings to each other. Things that are useful, edible, poisonous, or dangerous all need to be named—and remembered. The long apprenticeship of childhood prepares us for becoming an effective member of our group in a complex and many-faceted world. To deal with our natural and social surroundings we must name and categorize its many aspects. Most adults have a keen comparative eye for differentiating the creatures that might threaten them from those that are more benign. In addition, we categorize many things around us so we can deal appropriately with them or speak about them. What I didn't realize was how deeply innate this ability is, or how early it expresses itself. Visiting the lion house at the zoo with my daughters, then aged four and six, I had the opportunity to check out their ability to classify animals. As we stood in front of the cage of a fine black leopard pacing back and forth, I asked the girls whether this leopard looked like my sister's black Labrador retriever (Blackie) or her gray-striped cat (Tommy). The girls' immediate and unified response was "He looks like Tommy." Here was an opportunity to examine their response.

Feigning surprise and dismay at their decision, I announced that if he was big like Blackie, and if he was black like Blackie, then surely the leopard must resemble the Labrador retriever. "No, no, no!" came their agitated reaction. I persisted: "What makes him look like Tommy?" Their reply was simple: "He looks like a cat." Not satisfied with so superficial an assessment, I insisted on more facts. "His face, his ears, his tail" proclaimed the girls. But I continued to insist that the black leopard resembled my sister's big black dog. Finally, the younger girl, quite exasperated with her father's stupidity, declared: "Look! He walks like a cat!" Forced to admit my error, I did so. The girls marched off to see the lions, never doubting their original assessment. I've often thought of that little episode; the size and color of the animal were irrelevant to a set of characteristics the girls had used to develop their mental category: they *knew* the leopard was a cat.

The business of natural science has taken our innate ability to name and categorize all manner of objects and applied it to everything in the universe. Beginning in the folk taxonomies of cultures all over the globe, people have exhibited a desire to catalog and understand the elements of their surroundings. Modern science has amassed millions of specimens in museums and reams of descriptive literature in libraries to document the diversity of nature. At first it was all a jumble, with new plants and animals being described from all around the world. Zoos featured strange animals and botanical gardens displayed exotic flora, while naturalists were listing an ever-growing inventory of nature's creation.

Late in his life, the German poet Johann Wolfgang von Goethe is quoted as having said that Carl Linnaeus's classification of flowering plants (*Species Plantarum*) was one of the most significant texts in his life. At first, I couldn't understand why a great poet would have found such a drab dictionary-like compendium to be so meaningful. Then it dawned on me: before Linnaeus, there was no consistent way of identifying the rich diversity of flowering plants. Linnaeus's system provided a simple way of categorizing and finding the names of many plants. Though utterly artificial, his system of classifying flowering plants by the numbers of their sexual floral parts was immensely practical. Related plants were sometimes placed in very disparate categories, and Linnaeus knew it; more natural systems of classification followed quickly. What Linnaeus had done was make order out of chaos, a first step in organizing and understanding the world's rich diversity of plants and animals. Since then, taxonomy, the business of naming and classifying plants and animals, has played a central role in enumerating the world's living and extinct diversity, and providing the database for our understanding of biological relationships and evolutionary trends.

## CLASSIFICATION AND EVOLUTION

The careful naming and classifying of plants and animals, as pioneered by Linnaeus and other early naturalists, has helped make the natural world comprehensible. The Linnaean hierarchy is a system of "nested sets," beginning with the species. It places similar species in genera and similar genera in families, with the hierarchy moving up through orders into classes, phyla, and, finally, kingdoms. We humans find ourselves in a genus with only one living species (*Homo sapiens*) grouped together with the great apes as a family, within the primate order (monkeys and their relatives), which, in turn, belong to the class Mammalia, in the subphylum Vertebrata of the phylum Chordata. Each one of these higher ranks represents a more inclusive set of creatures within the kingdom Animalia. Though there are many arbitrary steps in deciding where to draw the lines between genera, families, or orders, this is a hierarchical system that has proven immensely useful.

The Linnaean system was an attempt to inventory nature's diversity, characterizing many natural groupings of plants and animals. Jean-Baptiste Lamarck, the "father" of evolution, suggested that these natural assemblages shared a common ancestry and had differentiated over geological time. Darwin's and Wallace's theory, claiming that natural selection was the primary driving force in evolution, did not require abandoning the classifications in use; rather, it helped explain how a deep history produced the similarities we use in classification. The Linnaean hierarchies have, in general, proven to be concordant with a fossil record of 500 million years; and they continue to be effective ways of categorizing life on Earth. Interestingly, those historical divisions that have been torn apart by modern molecular research are among the most ancient and the most minute. As we've noted: blue-green algae are now bacteria, fungi are no longer plants, and the Archaebacteria are a newly recognized kingdom of life forms.

Classification has often been a contentious enterprise among taxonomists who follow different approaches or methodologies. Taxonomy focuses on species recognition, classification, and nomenclature. Systematics is essentially the same thing, using a wide array of data sets to unravel evolutionary history; phylogenetics is similar, but with a stronger focus on the actual pattern of diversification through time. Now that there are gobs of new data from DNA analysis, it is possible to compare the actual sequence of nucleotides on the same gene in diverse lineages, ranging from individuals to species, genera, and kingdoms. How this avalanche of confusingly variable molecular information can be best analyzed is a focus of current attention.

The currently popular methodology of classification, cladistics, claims that significant new characteristics come into existence at the speciation event, when two new species come into being as the original progenitor species disappears (by definition). Although not especially realistic, this scenario does produce clearly defined "evolutionary trees" that are easy to code and compare. Another current and popular evolutionary idea, the theory of punctuated equilibrium, claims that evolution occurs in sudden bursts of speciation between long periods of boring stasis. These two paradigms (cladistics and punctuated equilibrium) reinforce each other in the notion that evolutionary change occurs primarily during speciation events, something that has never been clearly documented. While such evolutionary trajectories undoubtedly have occurred in the history of life, there is no reason to believe that these popular new ideas have erased the contribution of gradual evolutionary change from our understanding of evolutionary mechanisms. We shall revisit the question of "punctuation versus gradualism" when we consider the last two million years of our own history.

Developed more than two hundred years ago, the Linnaean system of ranks or nested sets has served us well, both for information retrieval, broadly comparative surveys, and communication. Animals and plants that were classified as very disparate orders and classes have indeed had long and independent fossil records. Because the larger sets represent levels of greater morphological divergence, they also represent longer periods of separation. All this assumes that divergence increases gradually in time. (Mostly it does; but sometimes, as in "living fossils," it doesn't.) All told, our taxonomies and hierarchies have given us a pragmatic way of representing the patterns of evolution in the fossil record, as well as dealing with the living world around us.[29]

## INCREASING DIVERSITY THROUGH TIME

Comparing the histories of diverse lineages through the geological record has allowed us to evaluate more general trends, such as an increase in numbers and complexity over time. Questions of origination, extinction, and diversification rates can be compared across many genera or families of plants and animals, based on the accumulated taxonomies of the fossil record; all are neatly arranged according to the clearly ranked Linnaean hierarchy. Comparing widely separated lineages over geological time or across today's world can also give us important insights into more general trends and geographic patterns. The evolution of animal diversity through time is a good example. By making a detailed survey of listings of families of marine animals through geological time, Jack Sepkoski was able to plot the rise and fall of marine diversity over hundreds of millions of years. His analyses clearly showed the effects of the major extinction events, but they also indicate a history of increasing numbers over time.[30] Trends of increasing species numbers over time are especially dramatic in terrestrial biotas. We've already discussed how vascular plants have expanded their numbers over time, and the importance of flowering plants in supporting today's species richness on land, but other groups of organisms have also had the effect of creating a richer world.

Birds and mammals are an interesting example. Because of their larger brains and complex behavior patterns, young birds and mammals are simply too "expensive" to be allowed to go it alone. Unlike most reptiles and amphibians, birds and mammals are cared for by their parents until they are nearly fully grown. A little lion doesn't have the problem of having to catch rabbits; the pride will provide food until the cub reaches its mature size. When a young eagle leaves its nest, it is almost the same size as its parents and ready to hunt the same prey. What this means is that both lions and eagles have narrowly defined hunting niches, which leaves room for smaller cats and hawks to make their livelihoods in the same landscape. Dinosaur communities in the Cretaceous were not especially rich in species. It seems likely that a little tyrannosaur had to pursue little prey and gradually move up through larger prey before achieving the fearsome stature of an adult. Parental investment, so characteristic of birds and mammals, has allowed the "packing" of more species into the same habitat.

As we saw earlier, minor disruptions keep the rain forest in continuous flux, which helps sustain so great a number of species. Predators, parasites, and pathogens can act as especially effective "disruptive forces." Larger predators, minute parasites, and microscopic pathogens can all reach their victims more easily as their hosts become common and more crowded. This process regularly decimates the more dominant species, making room for others, allowing habitats to support a greater variety of species. A surprising discovery, made only a few years ago, has found that ocean water is teeming with viruses. These submicroscopic beasties are very difficult to collect and identify, which is why they had not been noticed earlier in ocean water. Subsequent experiments, with tanks of virus-free fish and plankton in cleansed sea water, were expected to show an increase in biodiversity. Not so! As often happens in our most illuminating experiments, the results were exactly the opposite of what had been anticipated. Both biomass and species numbers plummeted in the super-clean experiment. Not only do the viruses keep the system from being dominated by fewer species, they also keep the system churning by allowing nutrients once locked up within their hosts to be recycled more rapidly.[31] Here, nature provides us with another example of how a "bad" cause (viral disease) can produce "good" results (sustain greater diversity). Though we may not like to admit it, stealthy predators and deadly pathogens are another part of the explanation for why there are so many kinds of plants and animals in our world.

Not only have the many factors we've explored in this chapter conspired to give us biotas so rich in diversity, these same processes seem to add more variety to the world over the course of evolutionary time. Historical trends of increasing numbers, greater morphological complexity, and more diverse behavioral repertoires are often interpreted as evolutionary progress. However, there are a few scholars who view the notion of evolutionary progress as a vestige of hopelessly outdated nineteenth-century ideology. At a symposium on the concept of progress in evolution held at Chicago's Field Museum, Stephen Jay Gould declared: "Progress is a noxious, culturally imbedded, untestable, nonoperational, intractable idea that must be replaced if we wish to understand history."[32] This is not ordinary scientific rhetoric. What's this all about? Having chewed on various aspects of biological diversification, let's move next to the question of "progress" in evolution.

# 6

# RANDOM WALKS OR PROGRESSIVE TRAJECTORIES?

clever species

SETI, the search for extraterrestrial intelligence, is based on the notion that other advanced technological societies should be fairly common in our galaxy. This idea, in turn, assumes that both biological and cultural evolution are progressive. However, progress is a subject that is rarely encountered in the recent literature of biological science. A comingling of the concepts of progress and purpose has encouraged scientists to avoid the concept of progress. Over the last two centuries, and for many people today, the notion of evolutionary progress has been closely associated with a planned and majestically unfolding creation, under the supervision of a paternalistic Creator. Divine purpose and miraculous actions, however, do not lend themselves to being tested or objectively verified, and they must therefore remain outside the playing fields of scientific discourse. But must we also banish the concept of progressive change from our discussions?

What a few hypercritical biologists seem to forget is that the word *progress* has an array of meanings. Doctors commonly speak of how diseases progress. And if diseases can progress, it seems to me that the history of life can progress. Surely, a land biota decorated with cloud forests and acacia savannas, where once there were only puddles of blue-green slop, is a dramatic instance of progress—progress in numbers and progress in levels of morphological complexity. In discussing Darwin's impact, Thomas Kuhn wrote: "Natural selection, operating in the given environment and with the actual organisms presently at hand, was responsible for the gradual but steady emergence of more elaborate, further articulated, and vastly more specialized organisms."[1] This phrasing, I believe, characterizes what most of us are willing to call evolutionary progress.

If we use the word *progress* in a loosely defined sense, devoid of plan, purpose, or predestination, we'll find progress almost everywhere. And why not? Ours is a universe of one-dimensional time. Almost everything changes with time, and time travels in only one direction. In the world of geology, there may be trends of a cyclical nature, but in the biological world, strong directional changes have occurred over time. "Indeed, the series of morphological and physiological innovations that have occurred in the course of evolution can hardly be described as anything but progress," noted Ernst Mayr.[2] "Man may not be the measure of all things," wrote Julian Huxley, "but the difference between man and the simplest organisms is certainly the measure of biological progress."[3]

George Gaylord Simpson devoted a chapter to the concept in his book *The Meaning of Evolution*.[4] However, *progress* is a word rarely seen in the contemporary literature of evolutionary biology. "Progress and superiority have no place at all in an objective evaluation of the vicissitudes of life's history," wrote Niles Eldredge in *Life Pulse*.[5] It certainly is true that extinction events have reduced the diversity of life, but after each extinction event, living things diversify anew, invent new modalities, and *progress*. They may be decimated all over again, but it doesn't take too long before nature, like the mythical Greek Sisyphus, has the ball rolling uphill yet again. And then there is also the "progressive" trend of increasing complexity and structural variety.

Let's admit that there are some striking exceptions. Some life-forms, like the ginkgo tree, the horseshoe crab, the lobe-finned fish *Latimeria*, and the American possum seem to have changed very little over tens to hundreds of millions of years. Bacteria have changed their external appearance not at all over several thousand million years. Clearly, "progress" is not an inherent drive in the evolution of living things. Nevertheless, life seems to have a tendency to elaborate its attributes and diversify. For example, Cambrian crustaceans had about a third of the morphological diversity of today's members of this phylum (shrimps, crabs, isopods, and relatives). By the end of the Carboniferous (280 million years ago), Matthew Wills estimates that these creatures had reached four-fifths of today's disparity, and he concludes that, once occupied, extremes of morphological form "tend not to fall vacant again."[6] Likewise, a recent study of 588 genera of ammonoids (fossil cephalopods resembling the chambered nautilus) covering a period of 140 million years, found that descendants were twice as likely to be more complex than their ancestors.[7] Extinctions turned the tables on many of the more complex ammonoid lineages, but the critters then began getting more complex once again. This surely looks like progress.

Michael Ruse claims that a strongly progressive philosophy of human history, developed in the eighteenth and nineteenth centuries, has both inspired and debased evolutionary biology from its earliest beginnings. In an otherwise fine survey of evolutionary biology over the last two hundred years, Ruse seems to view the concept of progress as if it were a virus that continues to contaminate and debilitate our science. Moreover, in commenting on various measures of progress, he remarks, "Complexity is a poor guide since ancient forms (trilobites) were more complex even possibly than humans."[8] Trilobites may have had lots more legs and body segments than we do, but one glance at the nervous systems will make clear who's way out in front. Stephen Jay Gould goes even further in his book *Full House*, arguing that "progress does not pervade or even meaningfully mark the history of life," and that "progress is . . . a delusion based on social prejudice and psychological hope."[9] He sees increasing complexity as nothing more than random and increasing variability expanding away from its simplest beginnings. Gould is perfectly correct when he insists that there is no evidence for natural selection having an inherent progressive drive. However, he fails to consider the possibility of higher-level environmental settings that might propel progressive change. Such "emergent" trends may characterize only a small percentage of evolutionary trajectories, but that does not deny their existence *or their importance*. If Gould were correct, then SETI is an even more futile effort than most of its critics have maintained. If "progressive trends" are not commonplace in nature, then the probability of contacting other extraterrestrial civilizations seems pretty bleak. However, progress *is evident* throughout the fossil record.

## PROGRESSIVE EVOLUTION

An interesting progressive pattern seen in some fossil animal lineages, especially fossil mammals and dinosaurs, is that species within a lineage often become larger over geological time (Cope's Rule). Though this generalization has clear exceptions (such as rats and mice among mammals, and birds among dinosaurs), this does seem to be a common trend over geological time.[10] There may be several reasons for such trends among animals. For one thing, many new lineages start off small. With lots more little species than big ones, it's a statistical likelihood that innovative lineages will arise from smaller beginnings. Then there's competition among males for access to females, where greater physical size usually has a clear advantage. Female choice may also select for larger males over a wide array of animals. In a study of the preferences of female bullfrogs, scientists statistically demonstrated that the biggest males were the most attractive. Unfortunately, larger organisms often have higher extinction rates than their smaller brethren, and many soon drop out of the geologic record, with the result that such trends toward larger sizes have not played a significant role in arguments about evolutionary progress. For this discussion, let's shift our perspective to broader trends including plants as well as animals.

In the previous chapter we argued that much speciation is driven by selection for ecological differentiation, helping explain why nature has created a planet so rich in species. Here we ask the question: how can natural selection produce "better adapted" species or more complex levels of organization? We've got to be careful with value-loaded words like *better*, and make clear the criteria we intend to use. Surely erect green plants are a more impressive solution to the problem of living on land than a thin scum of blue-green bacteria. The bacteria may be "better" in their ability to survive drastic changes in the environment. But when it comes to structural diversity and providing the basis for a richly three-dimensional land vegetation, bacteria leave a lot to be desired. As the eons have passed, the biological world has become richer in numbers, richer in the diversity of elaborate structural forms, and richer in the number of complex organisms capable of living under severe environments. Bacteria may display a huge diversity of biochemical pathways in their varied lineages, and they may be able to survive extremely severe conditions, but they are tightly constrained in their morphology. There is no such thing as a larger organism or tissue made up of bacterial cells. Nucleated cells marked a major advance in the history of life; complex animals were another quantum leap forward. Jellyfish are composed of seven different types of cells, worms about forty-two, insects about sixty, and humans over two hundred. Interestingly, plants, animals, and fungi devised their larger, more complex forms *independently*; each group constructs its larger morphology in very different ways. Just as significantly, these increases in complexity did not leap suddenly into being; they have accrued slowly over time.

Perhaps it is important to pause here and remind ourselves that no bacterium, fungus, plant, or animal has ever really evolved or, for that matter, progressed. Each of us is stuck with the hereditary information our parents gave us, and this doesn't change much. What actually *evolves* is the genetic information *of a population or a species, as gene frequencies change from generation to generation through time*. These shifts in gene frequencies may be driven and constrained by natural selection. But chance losses, population crashes resulting in genetic bottlenecks, geographic founder events (where populations are initiated by a few individuals carrying only a small fraction of the species' genetic diversity), and other effects can also have important consequences. Populations evolve over time as they gain new mutations and lose old ones. Though mutation is essentially random, the requirement for viability and reproductive success is a ruthless filter that will screen out the vast majority of new mutants. (The likelihood of a new mutational change being an improvement over a genetic heritage that has already proven successful over many thousands of generations is minuscule.) The necessity for a continued integrity of development and organization is an especially severe constraint on the amplitude of change possible in the evolution of more complex plants and animals. Quantum leaps to new levels of organization are extremely rare. And when they do occur, they often result in constraints that cannot be undone. The earliest land vertebrates had only four legs and we've been four-limbed ever since. Insects with six legs, spiders with eight, and crabs with ten, are other examples of ancient phyla having retained a fundamental body plan unchanged for hundreds of millions of years. All of us mammals, from giraffes to cats and bats, have seven vertebrae in our necks—without exception! A basic body plan, once developmentally integrated, is almost impossible to change (we call this phylogenetic constraint).

"When there are hidden strata of stability, one above the other, as there are in our universe, it follows that the direction of time is given by the evolutionary process that climbs them one by one," wrote Jacob Bronowski in 1970.[11] "What evolution does is to give the arrow of time a barb that stops it from running backward; and once it has this barb, the chance play of error will take it forward of itself." Once a new plateau of complexity has been reached, such as the nucleated cell, it provides a higher level of stratified stability that is unlikely to suffer reversal. (Actually, a few oddball organisms, mostly parasites, have greatly simplified themselves, but they do not negate the more general trends.) Each new level of complexity, in turn, provides the basis for developing yet more elaborate levels of organization, "some of which will chance to be stable." According to Bronowski, the increase of overall biological complexity over geological time is inevitable.

Admitting to a feminist viewpoint emphasizing cooperative organization rather than competition, primatologist Alison Jolly writes: "The greater transitions in evolution have produced ever larger matrices of cooperation. In this sense there has been a genuine directionality to evolution, toward more and more complex interconnections."[12] Such matrices began with closely interacting molecules within the earliest cells; endosymbiosis helped build the more complex nucleated cell, and these cells later formed cooperative networks to create plants and animals. Similarly, Lynn Margulis sees symbioses as a continuing force for the creation of new species and increased "complexification." Finally, complex ecosystems and elaborate social systems have carried these trends to ever higher levels of interactivity. Human evolution, she argues, appears to be following similar dynamics as our cultures become more interconnected and more interdependent.[13]

In his short book *The Basis of Progressive Evolution*, G. Ledyard Stebbins writes that the "more highly organized groups of animals and plants did not evolve through progressive adjustments of large populations living continuously in the same environment, . . . [T]hey evolved by more roundabout methods. . . . If there had not been periodic opportunities for unusual, exceptional genotypes to dominate their own small populations and radiate into marginal habitats, major evolutionary advance would never have been made."[14] Stebbins's view is reminiscent of Ernst Mayr's proposition that peripherally isolated populations are the most likely source of new species. Both views imply that significant new genetic configurations are more probable within smaller populations living in peripheral or more stressful environments.

The conquest of the land by the first green plants required the kind of progressive steps to new levels of organization that Stebbins had in mind. The spore-dispersing little liverwort was a really major shift into a new and different environment. With the further advance of internal plumbing, vascular land plants reached another stage where adaptive diversification began an expanding trajectory. Richard Fortey remarks: "It is almost impossible to describe the sequences of changes that happened in early [land] plants without using the language of 'improvement.'"[15] In the next chapter we will focus on the evolution of intelligent terrestrial animals, but it is doubtful that such creatures could have evolved without a rich and structurally complex land flora to sustain them. Ian Stewart and Jack Cohen interpret this as an example of what they call "complicity," where two or more complex systems interact, leading to new behaviors not present independently in the systems. They see evolution as a universal process, "a general mechanism whereby systems can spontaneously become more complex, more organized, more startling in their abilities."[16]

Life began and diversified in a warm liquid medium; the ability to live on land and under extremely dry or cold conditions came much later in evolutionary time. After having mastered the art of living on land, plants and animals slowly proceeded to expand from moist tropical lowlands into ever more inhospitable habitats. When you see palm trees growing outdoors you know you are in the tropics or subtropics; palms have never learned how to deal with really cold temperatures. Arid deserts and polar regions may have had bacterial residents for a very long time, but these were the last areas to be invaded by more complex plants and animals. Penguins and polar bears hunt from ice floes that had no comparable passengers 100 million years ago. What might be the forces driving such trends?

The most reasonable selective forces underlying such general directional trends are the effects of predation and competition. In a dangerous world, peripheral environments may be physically more severe but free of disease and competition. John Bonner calls this the "pioneering effect," and he sees it as the reason for the invasion of ever more severe habitats, such as the coldest polar regions and the darkest ocean depths.[17] This scenario claims that competition (whether between different species or between individuals of the same species), together with disease and parasitism, has been the driving force behind trends that have populated some of the world's most severe habitats. These insights reveal a serious flaw in many current discussions regarding the possibilities of life elsewhere in our solar system or beyond. Such discussions assume that if bacteria can live in near-boiling water, in rock fissures deep underground or within frigid Antarctic ice, then it is reasonable to expect life-forms in similar environments elsewhere in the universe. Hogwash! The flaw in these assumptions is that they ignore the problems of life's origin and earliest evolution. Bacteria living in extreme environments today are the product of thousands of millions of years of intense selection; it seems highly unlikely that they also *originated* in such environments. Meanwhile, competitive forces have also been sculpting life in less severe environments, driving a rich variety of evolutionary trajectories. One of the more interesting effects of these trends is biological mimicry.

## MIMICRY AND PROTECTIVE COLORATION

Predation, competition, and herbivory have influenced animal and plant evolution in many different ways. Visually, one of the most fascinating results of these competitive interactions is the phenomenon of protective coloration. Camouflage is the most obvious example, but brilliant colors are also part of the story. The unpalatable viceroy butterfly has strongly outlined black veins on a bright orange background, just like the more common and more poisonous monarch butterfly. These brightly contrasting colors serve as a warning signal to those who would eat them. If every bird had to learn for itself that monarchs or viceroys were the wrong place to look for dinner, these butterflies would soon be extinct. Apparently, over time, birds have become genetically programed to avoid these brilliantly colored butterflies; by closely mimicking each other, the butterflies retain a clear and consistent signal for those who might otherwise devour them.

Camouflage or color patterns that mimic their surroundings allow different species of wolf spiders to hunt in sand, grass, or dark leaf litter with less chance of being detected by potential prey. Caterpillars resembling bird droppings and moths looking like dead leaves or bark mimic specific aspects of their surroundings to avoid keen-eyed birds. One of the finest examples of recent short-term evolution comes from population changes in the British peppered moth (*Biston betularia*). Collectors of lepidoptera had paid higher prices for the rare dark form (forma *carbonaria*) of this moth in the early 1800s. But as the industrial revolution progressed, the dark form of the moth became common in areas of heavy industry, and its price tag dropped. Air pollution had two effects: sulfur dioxide killed the grayish lichens found on tree bark, while carbon soot made the grayish bark and leaf litter darker. Experiments with birds confirmed what everyone had surmised: the resting moth matching the color of bark was less likely to be found and eaten. Recent improvements in air quality have caused tree trunks to return to their natural paler hues, and today the great majority of moths of this species are grayish once again. (This classic story has recently been challenged by a critical review of earlier research, but I suspect the original interpretation is essentially correct.)[18]

One of the most often illustrated cases of protective form and color is that of certain tropical broad-winged grasshoppers and butterflies mimicking a living or dead leaf as they sit motionless on a slender twig. Some of these creatures are quite impressive. They are not just green leaves, but green leaves with a midvein, small brown blotches, slightly tattered edges or other realistic details. A biologist from Brazil, visiting Chicago's Field Museum some years ago, brought with him a cigar box holding about twenty such deceased leaf-mimicking grasshopper specimens. Neatly pinned in rows, all belonged to the same species and were about the same size as our katydids, with bodies two inches (5 cm) long. But they were quite varied. One or two were brown, a few were entirely green, and the rest were green variously decorated with brown or gray markings. The point of showing us this box was important; our Brazilian colleague had raised all these insects from a single batch of eggs. All were the offspring of the same mother! Here we could plainly see the rich palette of variability being maintained in a local population. Just as the peppered moths had maintained dark variants in their population and still do, high levels of variability among the grasshoppers enhance the chances for the species' survival in a complex and unpredictable environment. (We humans aren't much different, about 85 percent of our genetic variation is found *within* populations, helping maintain high levels of local variability.)

Even plants get into the business of mimicry. A few species have leaves bearing little rounded white projections that look like insect eggs. Apparently, that's exactly what they're supposed to look like. The *reason* these plants do this is tied to the behavior of the butterflies whose caterpillars feed on these little weedy plants. The plants, of course, haven't figured this out; it is probably a small morphological feature that came along accidentally and is maintained by natural selection. A crucial point in this story is the nasty feeding habits of the caterpillars themselves: the bigger caterpillars will eat more than just the leaves; given the opportunity, they'll eat smaller members of their own species. This has resulted in an unusual behavior pattern: the egg-laying butterfly will first check the plant for eggs already deposited by a previous butterfly, *before* laying her own. I never thought that butterflies could be this smart. Again, they probably don't think about it; this is a genetically based inborn behavior pattern. By carefully removing the fake eggs from marked plants, a European investigator tested the "egg-mimicry" interpretation. Sure enough, there was a statistically significant increase in the number of real eggs laid by the butterflies on the "treated" plants, in contrast to the ungroomed control plants that still possessed all their dummy eggs.[19]

Let's take a little diversion here. The little white dummy eggs bring up another interesting problem in biology. Naturalists are often eager to infer functions for complex structures well before having any real data as to what the structure or device might do; this is called the adaptationist paradigm. The underlying notion is simple; if nature has crafted an intricate biological device, that device ought to be doing something useful. Nature is economical: energy isn't wasted on nonperforming structures. The dummy eggs require energy to construct; and where there is a cost, there better be a benefit. These dummy eggs are decoys, contributing to plant survivorship. More fundamentally, there's a simple reason to imply function for fancy biological structures—the great majority of mutations are *bad*. An incessant drizzle of deleterious mutations will erode the genetic underpinnings for any structure unless there is ongoing selection for the continued function of that device (or it is tightly linked to a functional trait). In the real world of ongoing severe selection, the message is simple: use it or lose it.[20]

There are other kinds of deceptive strategies and mimicry in the world of flowers. A few species of large rain forest *Cydista* trees (members of the catalpa or trumpet creeper family) produce a huge number of colorful flowers over a period of only a few days. These usually sparsely distributed trees are quite spectacular in the rain forest as they suddenly become a blaze of color amid a sea of greenery. But there's a catch. The flowers produce no nectar; there's no reward for the busy pollinators. These species rely on insects that will visit them as they visit various other flowers. Significant pollination is effected before the insects "wise up" and go elsewhere for nectar. This is called pollination by deception and, like most other criminal activities, is quite rare. These nectarless species are uncommon; they are mimicking the many more tree species offering ample rewards to those who serve them.

More than any other plant family, orchids demonstrate a great variety of deceptive patterns. But why might orchids produce many more deceitful species than other plant families? The answer seems to be that, unlike most flowers, orchids produce pollen not as loose powdery grains but in special packets (the pollinia) containing thousands of minute pollen grains. In orchids, a single pollination event can produce thousands of seeds. Admittedly, these are minute seeds that need to germinate in just the right spot to create a new plantlet, but the seeds number about two orders of magnitude greater than for most other plants. What this means is that orchids don't have to win often, because when they do win (get pollinated) they win big. Not only do quite a number of orchids, such as our lovely lady slippers, produce no nectar for their pollinators, some species have developed a truly bizarre form of mimicry. These strange-looking little orchid flowers are so similar in appearance to the bodies of specific insect species that the male insects attempt to mate with the flower! Making this mistake more than once will result in having the myopic male carry pollinia from one orchid to another—and produce tens of thousands of seeds. And it's not just the shape of the flower that inspires the passion of these males; a recent study of an insect-mimicking *Ophrys* orchid disclosed a mix of aromatic chemicals virtually identical to the aromatic pheromone mix of the female wasp; here's a case of chemical mimicry.[21] This form of pollination, called pseudo-copulation, is found in no other family of flowering plants, probably because it is a strategy that doesn't work very often. Orchids get away with it because they are the only plant family capable of producing so large a number of seeds with only a single pollination event.

Mimicry and protective coloration provide ample evidence for the reality of strong selective pressures created by a hostile environment of predators and pests. However, mimicry may not convince the skeptical observer that a truly "progressive" trajectory has taken place. For that we need to look at coevolution and arms races.

## COEVOLUTION

Continuing with orchids, a dramatic example of coevolutionary interaction has resulted in an orchid, *Angraecum sesquipedale* native to Madagascar, having an eight-inch-long spur. The nectar-holding spur is a narrow tubelike structure at the back of the flower. When Darwin heard about the discovery of this orchid, he quickly postulated that there was a hawkmoth, the orchid's probable pollinator, with a tongue equally long. That moth was soon collected, but it was more than a hundred years later that such a moth was actually observed pollinating this orchid in the darkness of night. But why did both spur and tongue become so long? Darwin assumed that the orchid is always under selective pressure to position its pollinia on the front of the moth where it can be carried to another orchid for pollination. Without such pollination and consequent seed production, the orchid species would go extinct. On the other side of the relationship, there is a contrary selective pressure for the hawkmoth to stay clear of the orchid so that its flight is not interrupted. Like hummingbirds, hawkmoths feed while hovering with extremely fast wing beats. Coming into contact with parts of the flower can damage their delicate wings or disrupt their flight. Thus, the eight-inch spur and eight-inch tongue seem to be the result of the continuing interaction of two countervailing forces at play: contact for the orchid and avoiding contact for the moth. One can imagine how populations of both orchids and moths have been under continuous selective pressure, with both the nectar-containing spur and the hawkmoth's tongue slowly becoming longer over thousands of generations.[22] Similarly, a tropical American hawkmoth (*Amphimoea walkeri*) sports an even longer tongue (10 inches, 25 cm), and we know of a few white night-blooming flowers with narrow floral tubes about that long, such as *Posoqueria grandiflora* in the coffee family and *Tanaecium jaroba* in the trumpet-creeper family. Although actual nocturnal pollination has not been seen in these species, both insect and plants share the same range, from Nicaragua southward into the Amazon basin. This supports the idea that all are part of a coevolved pollination syndrome.

Though less closely correlated, insects and colorful flowers have been codependent and loosely coevolving for at least 100 million years. Honeybees are probably a later and more modern lineage to make use of this relationship. Honeybees gather two important resources from flowers: nectar for their own energy needs and both nectar and pollen to feed their young. Recent experiments have shown that honeybees not only recognize the difference between symmetrical and asymmetrical patterns, but they also remember the difference and prefer the symmetrical. Other insects probably have similar capacities and this may help to explain why so many flowers are so unvaryingly symmetrical: the bugs like it that way. Since colorful flowers are "using" insects to transfer pollen, it makes sense for natural selection to keep a tight control on the patterns and shapes that work best.

Birds, too, can be important pollinators. One of the most obvious examples of diffuse coevolution is the narrowly tubular red flowers found throughout the tropical and warm temperate regions of the Americas. These bright red flowers are usually held horizontally and positioned away from other flowers or leaves. There is no broad "landing field" at the mouth of these flowers and, if they have prominent petals, these are bent backward or downward, out of the way of the visiting pollinators. You will find no such flowers in the native vegetation of Africa, Asia, Australia, or Europe. The reason is simple: they have no hummingbirds. For birds, as for ourselves, red makes a strong contrast against a background of green foliage. Red is the characteristic color of flowers pollinated by birds throughout the world; but in the Old World there are usually places for birds to perch while sipping nectar. Hummingbirds have such rapid wing movements that they can remain stationary in flight or fly backward. Like the hawkmoths, they fly while feeding.

In the real world, full of nasty neighbors, it's really nice to have your own police force. A few species of acacia have done just that in a mutualistic relationship with ants. In Central America, you can often spot these small trees (*Acacia cornigera* and allies) from a distance; they've got only sparse vegetation beneath them, and there are no vines growing over them. If you grab hold of a twig, you'll quickly find out why, as fierce little ants rush out to sting you. These same ants girdle stems of vines and nearby plants, allowing the acacia more access to light. For all these services, the acacia produces protein bodies (beltian bodies) at the tips of leaves and, as if that weren't enough, special nectaries along the stems provide the ants with energy-rich sucrose. All this sounds like a great example of mutual benefits; but the bad news is that maintaining a small ant army takes a lot of energy, which may explain why only a few acacia species out of more than a thousand have got themselves tied up in this expensive symbiosis. But there are some energy savings. A careful comparison of the leaf chemistry of an ant acacia and its antless relatives in Mexico has shown that most of the ant-supporting species do not produce the "expensive" cyanogenic compounds that the other species synthesize to make their leaves less palatable.[23]

Unusual mutualisms and diverse pollination strategies have helped flowering plants become the most species-rich group of plants, providing resources for complementary evolutionary diversification in insects and other animals. When confronted with the question "What can a biologist say about the nature of God?" J. B. S. Haldane is said to have responded: "An inordinate fondness for beetles." Numbering well over 280,000 named species, beetles are a phenomenon to reckon with. A recent analysis of beetle diversification has shown that those lineages that adapted to feeding on flowering plants are generally far more diverse than those lineages that did not.[24]

## CONVERGENCE AND DIVERGENCE IN ECOSYSTEMS

Botanists have found that tropical rain forests and some seasonally dry forests seem to have much the same structure in far distant parts of the world. This seems to be true despite the fact that the different regions have floras made up of unrelated species and genera. Large leafless candelabra-like trees of the cactus family are found only in dry areas of the Americas. Trees of very similar form have also evolved in Africa, but they belong to the genus *Euphorbia* in a very different family. Sample surveys of small plots in similar environments have yielded similar numbers of plant species throughout the world. These observations led to the notion that, given the same conditions, different areas of the world would produce convergent biotas. For example, Mediterranean climates are found in California, central Chile, southwestern Australia, South Africa, and of course the Mediterranean region. All are located between 30 and 40 degrees of latitude (north or south) on the westward side of large continental areas. Their climates are very similar: a hot, dry summer and a dry fall are followed by a cool, wet winter and a moist spring. Since a great majority of the species bloom only in springtime, the floral displays in a Mediterranean-type vegetation can be spectacular at that time. There is also a strong similarity in the form and growth patterns of the plants in these vegetation types. Trees are short and tend to have small, tough, drought-resistant leaves. Overall, 40 to 50 percent of plant species in Mediterranean-type flora are short-lived annuals, and with such similarity of flora, there is often a similarity of animal life.

Convergence of vegetation in different parts of the world led zoologists to expect that animals would also show convergences. Some of these similarities may be numerical: the Mediterranean-type vegetation of southern California has 235 species of nonmarine birds, very similar to the 230 nonmarine bird species in Chile's Mediterranean vegetation. Like cacti and euphorbias, there are some examples of very similar body plans and lifestyles in animals on different continents. A number of Australian marsupial mammals are similar in general form and life habits to unrelated placental mammals in other parts of the world. These similarities include critters looking like moles, woodchucks, small dogs, and jumping mice. The pangolin of Africa resembles the armadillo of Latin America. Getting back to statistics, plots in the semideserts of the southwestern United States and central Australia show very similar numbers of seed-gathering ant species. But there are many other parts of the world where no such similarities have been found.

Reptiles and amphibians are not as numerous on the forest floor in Malaysian rain forests dominated by dipterocarp trees as they are in American rain forests, despite similar rainfall and forest structure. The reason seems to be that the Asian dipterocarp forests do something Latin American forests do not do: a majority of these trees fruit at the same time (called mass fruiting)—after a number of years during which they produced no fruit at all. Unusual weather of El Niño years may be the critical trigger that initiates reproductive synchrony among so many different species. Fruit and seed consumers can maintain only minimal populations during the long periods of scant fruit fall in these Malaysian forests. Animal consumers are unable to quickly increase their numbers when mass fruiting occurs, and they have little impact on the abundant fruit crop. Because fruits are usually produced regularly over the years, American rain forests are able to support a significantly larger population of reptiles and amphibians than do the Southeast Asian dipterocarp forests.

A variety of gliding animals are another unusual aspect of tropical rain forests in Southeast Asia. These include not just squirrel-like mammals with flaps of skin along their sides but frogs with expanded webbing between fingers and toes and lizards with thin lateral extensions on tail and legs. There is no similar concordance of gliding animals in other tropical rain forests. Rain forests in Southeast Asia are distinguished by tall trees to more than two hundred feet in height, creating a more open canopy; tropical American and African rain forest trees rarely grow to such heights. Perhaps this is the factor that has provided a hospitable environment for the evolution of so many gliding rain forest animals in Southeast Asia. Thus, while we do see some convergence in similar ecosystems on distant continents, there is a wide array of differences. Unusual initial conditions, unrelated lineages of plants and animals, and accidents of history have been confounding factors in all of these trajectories, counteracting any tendency toward deterministic environmental convergence. However, while not common, there are some clear examples of convergence. and these reassure us that some selective effects are, indeed, universal.

We can see those effects more clearly when we study faunas that shared a similar recent origin. A study of anole lizards on the four largest Caribbean islands (Cuba, Jamaica, Hispaniola, and Puerto Rico) have documented how these lizards evolved similar lifestyles independently. Grass/bush-inhabiting species evolved on three of the islands, while trunk/crown-inhabiting species evolved independently on all four islands. The authors of this study concluded that intense interspecific competition was the driving force behind these adaptive radiations, resulting in very similar anole faunas on all four islands.[25] Similarly, but more central to the question of biological progress, competition, disease, and predation can also be tightly species-specific in what is often called an evolutionary "arms race."

## EVOLUTIONARY ARMS RACES

In his influential book *The Blind Watchmaker*, Richard Dawkins wrote, "The arms-race idea remains by far the most satisfactory explanation for the existence of the advanced and complex machinery that animals and plants possess."[26] The arms race is a situation in which specific predator and prey species may be locked into a relationship that extends over long periods of time to produce important evolutionary consequences. Unfortunately, few contemporary biologists have specifically focused their research on the concept of progressive arms races. Paleontologist Geerat Vermeij is an exception. He has examined the question of what he calls "evolutionary escalation" by examining the rich fossil record of mollusks. Unlike most shell collectors, Vermeij has been especially interested in those specimens that were broken, damaged, or repaired during their lifetimes. Blind since childhood, Vermeij studies shells with his sensitive fingers. He became fascinated by the kinds of damage he encountered. By studying such shells in the fossil record and along today's ocean shores, he has attempted to gain an understanding of how mollusks have changed over geologic time. In his book on evolutionary escalation, Vermeij proposed several hypotheses to test the notion that competition and predation have played an important role in progressive evolutionary trends.[27] First, he hypothesized that competition and antipredator capacities of individual lineages have increased over time. Second, he suggested that recently evolved individuals should be better adapted to a more hazardous environment than their earlier ancestors. Third, as time proceeds forward, he predicted that hazards in the environment will have become more severe due to competition and predation. Vermeij tested these conjectures with examples from many sources, but damaged fossil seashells are his primary data set. As the crushing pincers of crabs have become larger and more powerful over millions of years, seashells—their prey—have developed thicker, stronger, and more elaborately ornamented shells. The survival of many seashell lineages with thinner shells, or crabs whose claws have not become enlarged, does not negate the fact that a few lineages clearly were part of an arms race, becoming more heavily armored (the shells) or more robustly powerful (the crabs) over the last 100 million years.

"Few things promote adaptation faster than becoming something else's lunch," declare Michael Majerus, William Amos, and Greg Hurst.[28] As in the case of the seashells, the strongest evidence for progressive evolution is that which has resulted from an escalating arms race between prey and predator or disease and host. The diminutive spaniel-sized ancestors of horses, running on five toes, were successful in their shrubby habitats of 50 million years ago. Today, much larger horses and zebras run faster and farther on the hoof (toenail) of a single digit. Their increase in size and speed was a trajectory driven by an increasingly dangerous environment of larger predators on open grasslands. Today, one cannot find little dog-sized antelopes on the African savanna grasslands where zebras roam. Little antelopes, like the dik-dik and klipspringer, are found only in rocky hillsides or in dense thornbush where they can find cover and where their small size is an advantage. Cheetahs, lions, leopards, wild dogs, jackals, hyenas, and a variety of smaller cats are a formidable assemblage that continuously culls sick and vulnerable grassland herbivores. It seems unlikely that herbivores or carnivores from millions of years ago would survive on Africa's open savanna grasslands today.

Pest pressure, like predator-prey dynamics, has produced some fascinating plant-animal interactions. Higher plants and animals have probably been "at war" in competitive arms races since their origins. Plants defend themselves with chemicals that include poisons, feeding deterrents, and compounds that block digestion. This leads us to the question of why plants might produce complex organic molecules to which we humans can become so strongly addicted—molecules that serve no purpose within the plant itself. The only reasonable explanation is that these compounds are chemical deterrents, discouraging browsing mammals from taking more than a few nibbles.[29] (Animals don't get addicted because they can't concentrate the drugs.) Also, on the animal side of the battle zone, some lineages have developed various mechanisms to detoxify, sequester, or bypass specific compounds that plants have developed for protection.

Fungi have also been part of an intensely competitive world, mostly beneath the soil. When driving past the Abbott Laboratories' production facilities north of Chicago on a windless day, one can often detect an aroma reminiscent of a deep forest interior. This is where the pharmaceutical company is busy culturing large batches of actinomycetes (fungilike prokaryotes), preparing to extract the antibiotics these organisms produce. They are especially abundant in the soils of moist forests and give the deep forest environment much of its characteristic aroma. Within those forest soils, many creatures have been at ceaseless war, competing for the same rich organic detritus that will sustain them. The reason for the production of effective antibacterial compounds, *our antibiotics*, is simple: it is the way actinomycetes keep bacteria out of their territory. The product of a long history of intensive competitive interactions, these compounds have become our primary armament in the battle against disease-causing bacteria.[30]

As in the case of the soil fungi and bacteria, an arms race between host and disease is not easily visualized; the host's defenses are largely internal and biochemical. However difficult to see, this interaction can have catastrophic consequences when a disease reaches an isolated host population that has not been part of the arms race. American chestnuts were devastated when a Eurasian chestnut disease was introduced into the United States early in the twentieth century. Once a dominant tree over much of the Appalachian mountains, hundreds of millions of chestnut trees were extirpated in less than fifty years. The stately American elm, whose tall spreading limbs graced so many of our older communities, has also been severely reduced by the ravages of Eurasian plant pathogens. Tragically, a comparable scenario devastated our own species, as Amerindian populations were suddenly confronted with human diseases from the Old World.

Archaeological evidence rarely provides precise figures for human population levels in ancient times. After contact with Europeans, estimates for overall native population decline in the New World range from 20 to 80 percent. Unusual evidence for a population crash has been uncovered by pollen studies in Veracruz, eastern Mexico.[31] Using a nine-meter core from the bottom of Lago Catemaco, scientists were able to analyze the pollen that had fallen into the lake over the last 3,000 years. Accumulating continuously over time, pollen leaves a record in the mud that slowly builds on a lake bottom. Resistant to decay, pollen can persist for millennia. At about 600 C.E., pollen from the Lago Catemaco core indicated that evergreen forests of the surrounding area had been cut down and replaced by maize agriculture. Maize and weed pollen then dominate the core for the ensuing one thousand years. Suddenly, at around 1600 C.E. the forest pollen returned to dominate the local region for the ensuing three hundred years. The pollen record bears witness to the fact that an area supporting maize cultivation for one thousand years reverted to forest shortly after Europeans arrived. Here, in one of the most dramatic events ever recorded by pollen analysis, is evidence for the sudden demise of an entire agricultural population.

Smallpox, influenza, measles, typhus, bubonic plague, diphtheria, cholera, and other maladies had interacted with humans for thousands of generations in Africa, Europe, and Asia. Since people of the Americas had not been part of these escalating arms races over the last ten thousand years, many lacked the immunities necessary to meet the challenge from the newly introduced and highly virulent pathogens. European contact with the New World initiated the largest population loss in human history.

Both Jared Diamond and Roy Porter argue that close interactions between peoples of the Old World and their domestic animals was the primary reason for the development of so many diseases there, and this view has some merit.[32] Influenza is a good example, regularly developing new variants by picking up bits of related viruses from domesticated pigs and geese. However, a much larger land area, many more people, and many more species of mammalian hosts seem to be the more general factors for the presence of so many more human diseases in Africa and Eurasia. The continuing generation of new diseases in the Old World has been highlighted by AIDS (acquired immune deficiency syndrome) caused by the human immunodeficiency virus (HIV). Closely related to similar viruses in chimpanzees and monkeys, the AIDS virus had its origin in equatorial West African forests where chimpanzees and monkeys are hunted for food. It stands to reason that

Old World monkeys and apes, being more closely related to us than any other major animal group, have been an important source of human pathogens. Nevertheless, from a more general perspective, both the population devastation suffered by Native Americans after European contact and the ravaging of chestnut and elm trees by Eurasian pathogens can be seen as strong support for Vermeij's third conjecture: evolutionary arms races have made the world a more difficult place in which to live.

## ESCALATING ECOSYSTEMS

Naturalists have speculated widely on just how natural selection works and how it might be measured. Biologists generally talk in terms of fitness: the ability of an individual to contribute genes to the next generation. By definition, an individual with greater fitness will have more offspring and contribute more genes to succeeding generations. An important thing to remember is that as the environment changes, so does fitness. A characteristic that may be detrimental during a period of wet years may have survival value during a dry year. Even though it may be very difficult to measure in nature, fitness is a useful concept. Some scholars have criticized the notion of fitness as being circular (a tautology): natural selection optimizes fitness and fitness measures natural selection. But selection is actually an algorithm of factors cadenced over time, and its effects have proven to be predictable, regardless of some circularity in the words we use to describe it. More important, there are other ways of looking at fitness and natural selection.

Leigh Van Valen of the University of Chicago approached the notion of fitness from a rather different perspective by thinking in terms of the species rather than the individual. He suggested that fitness might be a measure of an entire species' use of energy in the ecosystem in which it is living. And, he reasoned, if a species is increasing its overall fitness, it must be taking a greater bite out of the energy available in that ecosystem. We know that the energy of ecosystems is largely determined by the amount of sunlight falling on the system, as well as the energy stored in living and dead tissues of the system. Consequently, the amount of energy in any ecosystem is tightly constrained, ultimately dependent on solar energy. Thus, Van Valen argued, as a species increases its fitness it must affect all other organisms in the system *negatively*. Quoting the Red Queen in a story by Lewis Carroll, he suggested that each species finds itself in a world where "it takes all the running you can do, to keep in the same place."[33]

Van Valen analyzed extinction rates in the fossil record and was surprised to find that they seemed to vary little over time for each group he examined, discounting the major extinction episodes. This prompted him to propose what has become known as the "Red Queen hypothesis," an idea focusing on the importance of interactions between species living in the same environment. Other paleontologists had also noted a near-linearity in extinction rates within lineages, but they suggested that it was due to the random trials and tribulations of changing climates and other minor calamities. This view emphasizes being besieged by unpredictable changes in the physical aspects of the environment and has been characterized as the "Field of Bullets hypothesis." There is still considerable argument over whether extinction rates really are linear; perhaps they are an artifact of our taxonomies or analytic procedures. Regardless, both the Red Queen and the Field of Bullets are reasonable scenarios—and they are not mutually exclusive. Natural selection, random disasters large and small, and escalating arms races create a cruel and dangerous world. In fact, more detailed analyses of fossil marine organisms over Phanerozoic time by David Raup, Jack Sepkoski, and others indicates a slight but steady reduction of extinction rates within the same lineages over time, as if living things have been getting just a little bit smarter at the business of surviving through geological time.[34] Surely, that's one measure of progress.[35]

As we seek to understand evolving ecosystems, it may be useful to reconsider the underlying dynamics once more. Basically, Darwin's and Wallace's notion of natural selection is nothing more than common sense writ large: offspring differ, many of these differences are determined by hereditary factors, many of the offspring will die before they reproduce, and genes that promote survival and successful reproduction will be favored over time. Both Darwin and Wallace had been deeply influenced by Thomas Malthus's *Essay on Population*, in which the author argued that human populations would continue to grow geometrically while agricultural productivity could not.[36] From this straightforward analysis, Malthus predicted increased famine and disease among the soaring populations of Europe. Clearly, the world of nature had similar qualities: geometric reproductive potential in a world of limited resources. Natural selection simply expanded Malthus's idea, acting on inherited variability across a vastly broader horizon. Modern interpretations of evolutionary history support the theory that "selection" has played a powerful role in the historical development of lineages, as well as continuing to constrain the genetics of their contemporary morphology and physiology. Perhaps the best support for the significance of environmental selection is found in situations where some aspects of selection are relaxed. Historically, the most impressive animal radiations have taken place in environments of reduced competitive pressures, as in the early Cambrian, and after the great extinction events. It has been under these conditions of "open predator-free environments" where huge bursts of animal diversification have taken place. Unfortunately, for most of life's modern history, friendly habitats have been few and far between.

In today's world, some of the most dramatic examples of selective effects can be found in two of the world's most severe environments: deep caves and the deep ocean. Lacking light to provide a steady production of energy-rich foods, both deep caves and the sea floor are the equivalent of nutrient deserts. Under such conditions, severe ongoing selection has created extremely thin creatures that must survive on a minimal diet. However, the absence of light also eliminates selection for some traits that are essential in an illuminated world. Given enough time, cave animals become blind and colorless. Similarly thin and colorless animals are found on the dark floor of the deep oceans. In both kinds of habitats two factors seem to have been at play. Since deleterious mutations far exceed those that might be advantageous, mutations for loss of eyesight or loss of pigment production can be expected to manifest themselves early and relentlessly. In the blackness of a deep cave or the ocean floor, individuals lacking eyesight or pigmentation suffer no competitive disadvantage and they can flourish. Better yet, they are not wasting energy on attributes unimportant to their survival in these energy-poor environments. In such dark environments, there may be a clear advantage in no longer developing eyes or pigments.

Islands are another unusual environment in which some aspects of selection may be relaxed. Lacking predators, isolated islands have often produced a naive and vulnerable fauna. The Galápagos Islands have become a tourist mecca, thanks in large part to a friendly and fearless fauna. A number of islands in the Mediterranean Sea and the East Indies were the home of small, pony-sized elephants. Selection in an environment of limited food resources lacking major predators probably helped create such miniature pachyderms. Known only from their bones, these interesting animals vanished soon after human hunters reached the islands.

"Islands have been especially instructive because their limited area and their isolation combine to make patterns of evolution stand out starkly," writes David Quammen in his well-referenced book *The Song of the Dodo*.[37] Ranging from the role of islands in the creation of the concept of natural selection in the minds of both Darwin and Wallace, to the importance of Robert MacArthur and Edward O. Wilson's theory of island biogeography in contemporary conservation planning, Quammen's book surveys many aspects of island biology. He points out that island biotas have always been vulnerable to the introduction of exotic species, some of which have been able to explode in an environment of relaxed predation. Hawaii, the world's most isolated archipelago, is especially vulnerable to introduced species and is the home to a great many threatened endemic species. On the island of Guam, a brown snake, accidently introduced after World War II, caused a devastating decline in native birds and lizards. Interestingly, there are no spiny or poisonous plants in New Zealand's native flora; there weren't any herbivorous mammals around to chew on them (until introduced by humans). As in the examples we discussed of host-disease interactions and predator-prey relationships, the vulnerability of isolated islands is further evidence for Vermeij's conjecture that the larger world has become more dangerous with time.

The susceptibility of isolated oceanic islands to being ravaged by continental species is only one example of the fact that larger land masses have proven to be a more intensely competitive arena than smaller areas. A grand exchange of faunas took place when North and South America became united by a land bridge about three million years ago. Giant ground sloths, opossums, armadillos, and monkeys traveled north from South America. Deer, racoons, bears, wolves, as well as large and small cats, moved southward from North America. Armed with a fauna that ranged across most of the northern hemisphere, North America sent a phalanx of aggressive mammals into a continent that had been isolated for more than 50 million years. Two million years after the northern animals invaded, four endemic orders of mammals and many unusual genera and species that made the fauna of "island South America" so different were no longer a part of the living world.

Whether islands or continents, the results are the same: smaller land areas offer smaller stages for competitive interactions among species. Evolutionary patterns among the primates suggest that something similar has occurred over longer time periods among higher taxonomic ranks as well. Australia became an island before placental mammals were widely distributed. It is the only continent where primitive egg-laying mammals survive and marsupials dominate the fauna. There are only a few lineages of nonflying placental mammals in Australia, and there are no native primates. Madagascar separated from its southern neighbors when primates were at an early stage in their evolution, and the lemurs have flourished on that large island ever since. Today, this particular early lineage of primate evolution is found nowhere else in the world. In similar fashion, South America was an isolated "island continent" when monkeys arrived more than 30 million years ago. With three premolars on each side of the jaw, New World monkeys retain a more ancient condition than do Old World monkeys (and humans) with only two premolars. Also, their brain volumes are smaller than those of Old World monkeys of similar size, and most have color vision based on only two sensory pigments. Old World monkeys have three pigments of slightly different color sensitivity, allowing them (and us) to see a world of many different hues. For these reasons, Old World monkeys and apes are considered more advanced than their American cousins. Fossils provide convincing evidence that the most highly evolved primates, the Old World monkeys and apes, originated in Africa-Eurasia. But what happened to the early lemurlike primates of Africa-Eurasia, and why are there no primitive monkeys resembling the New World monkeys in Africa or Eurasia today? Fossil evidence indicates that their ancestors lived in these areas long ago. Were these more primitive lineages wiped out in a scenario of escalating competition, able to survive only on the islandlike land masses on which they are found today? It sure looks that way.

This chapter's title, "Random Walks or Progressive Trajectories?" presents a false dichotomy. Most evolutionary trajectories appear to be random walks as natural selection drags species along behind a constantly shifting and unpredictable environment. Such a pattern of adaptation is not what most people would call progressive. Also, selection will lead some species into narrow specializations that may prove to be blind alleys in which they are trapped. However, as was stressed previously, simple competition can give rise to niche differentiation and increasing species numbers over time. In addition to these general trends, there is the rare invention of a new level of molecular, physiological, or morphological complexity that may be the basis for a major new radiation. The development of the nucleated cell, complex animals, and land plants were the bases of such new radiations. Finally, there is the obvious fact that the environment includes many interacting neighbors, where adaptations in one lineage are likely to precipitate counteradaptations in other lineages. This gives rise to arms races and clearly emergent progressive trends. We are all loosely coevolving in richly diverse landscapes, while at the same time engaged in tight arms races with our specific parasites and pathogens.

The vulnerablity of island species to extinction makes clear that arms races have been most intense and prolonged on the world's largest contiguous land surfaces. And here again, it looks like we've been incredibly lucky with a just-about-perfect planet, providing land areas both large and small on which many different evolutionary dramas might unfold. Had the Earth's land surfaces been divided into many small areas, it seems unlikely that our own particular primate lineage would have reached its present state.

The take-home message here is simple. Just as Van Valen and Vermeij have argued, many aspects of the environment have become more difficult and dangerous with time, and it is on the Earth's largest land areas where progressive, competitive interactions have been the most severe. Here the single most significant example of progressive evolution was forged: the creation of a really powerful brain.

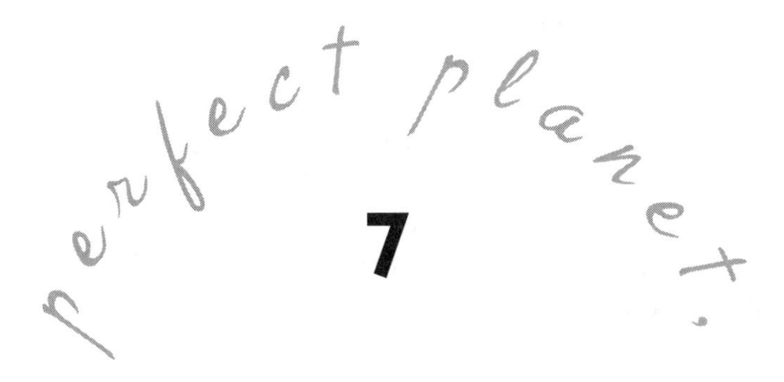


# 7

# THE ESCALATION OF INTELLIGENCE

clever species

When we talk about intelligence on planet Earth, we're talking mostly about creatures like ourselves, vertebrates with backbones. Except for the squid and octopus, invertebrates cluster at the bottom of the intelligence hierarchy. Vertebrates are different; neural crest cells develop early in the embryo and divide actively to create nerve cells, eyes, head structures, and bigger brains. In addition, jawed vertebrates have their nerve cells (neurons) encased in a fatty myelin sheath, insulating these cells from surrounding tissues and allowing nerve impulses to zip along their slender axons at fifty to one hundred meters per second. Thanks to their slender high-speed wiring, vertebrates were able to become large swift predators, nicely epitomized in an ancient lineage, the sharks and rays. However, when discussing higher intelligence, we must look beyond the sea.

The challenge of a terrestrial environment has raised intelligence to its highest levels. The really smart animals of the sea are whales, dolphins, seals, and their kin; all are mammals descended from the land. Perhaps the uniformity of the marine environment, or ancient body designs and lifestyles that worked so well for so long, explain why fishes and their invertebrate neighbors have not developed the larger, more versatile brains of their land-based relations. Interestingly, fishes with larger brains are those using an electric field to explore their dark or murky environs as they hone in on elusive prey.

Perhaps there is a corollary with diversity: the greater number of plant and animal species on land, as compared with those in the sea, is the product of a more varied environment. Far more unpredictable and more demanding, the land surface has provided a rich array of opportunities and challenges. The land environment was the setting in which two distinct groups of vertebrate animals developed bigger brains. Birds and mammals reached higher levels of relative brain volume—early in their histories and independently—while amphibians and reptiles advanced little beyond their aquatic predecessors.

Relative brain volume or "encephalization quotient" is a better measure of evolutionary trends than simple brain volume because it takes into account overall body size or body mass when making comparisons. In such comparisons, we find amphibians, reptiles, and most fish having brains only about one-fifth the volume of the brains in birds and mammals of similar weight. By graphing body weight versus brain weight along logarithmic scales, neurobiologist Harry Jerison has drawn polygons that depict the ratios found in a great variety of vertebrates. From these polygons it appears that fish, amphibians, and reptiles have been adhering to the same rules of brain formation for over 300 million years. In contrast, birds and mammals reside in graphed polygons that parallel—but are consistently above—the polygon for lower vertebrates.[1]

## SPRIGHTLY BIRDS, SENTIENT MAMMALS

In the case of birds, a number of factors may have been responsible for developing a relatively larger brain. The muscular and neurological challenge of sustained flight is likely to have created very strong selection pressures. Additional factors include life in treetops, walking on only two feet, nest building, and caring for the young. *Archaeopteryx*, the earliest feathered fossil, had a brain almost as large as that found in modern grouse, pheasants, and their allies. Today, the smallest brain among birds, relative to body size, is found in the largest species, the flightless ostrich. Interestingly, a few fast, bipedal predatory dinosaurs of the late Cretaceous period, similar in weight to the ostrich, had a brain of about the same size. As one might expect, the birds with the greatest learning ability (parrots, crows, and ravens) have the largest brains for their body weight.[2]

Having to fly through shrubs and trees gave birds a special challenge, and they responded with a high-speed visual system. But while they achieved higher brain volumes at the beginning of their history, birds seem to have been under no strong selection for further increase in brain capacity. With their forward appendages transformed into wings, birds created a very successful and comfortable niche for themselves. However, with those front appendages unavailable for other tasks, the airborne niche left little room for further development. The 150 million-year history of birds presents only a few examples of bigger brained lineages. Fortunately for us, mammals fared better.

As in the birds, mammals exhibited an escalation in relative brain size early in their history. Over 250 million years ago, before the dinosaurs became dominant, the largest terrestrial vertebrates were a diverse group called the mammal-like reptiles. It was from one of these lineages that the earliest mammals arose, about 200 million years ago. The fossil record cannot tell us if these early mouse-sized creatures were covered with hair or nursed their young, but the fossils clearly indicate a critical shifting of three small bones from the back of the jaw to become parts of the inner ear. (This scenario is reenacted during early development in the opossum, when the auditory chain of little bones is pulled free of the jawbone, with the inner ear becoming responsive to sound only in the sixth week after birth.) At the same time, early mammals became more sensitive to odors, resulting in the elaboration of the brain's neocortical region, which processes both auditory and olfactory stimuli. Jerison argues that mammals were using their newly refined senses to create a more accurate multidimensional picture of the world around them.[3] Then, with the rise of the swift and sharp-eyed carnivorous dinosaurs, the early mammals became creatures of the night.

Another early and critical change for the evolution of mammals was shifting to a metabolic rate higher than that of their reptilian progenitors.

The higher metabolic rate would help maintain warmer body temperatures, something especially important for animals that are both nocturnal and small. (Small animals lose heat more rapidly because smaller bodies have a proportionately larger surface area than more bulky bodies.) A secondary palate, separating mouth from nasal passages, allowed mammals to chew food while breathing, keeping a steady supply of oxygen coming into the lungs. Differentiated teeth and elaborate strengthening of the jaw made chewing more efficient. With only a single set of teeth at maturity, mammals developed precisely aligned upper and lower dentition, making cutting and grinding more effective. Changes in jaw anatomy and facial architecture also allowed for restructuring of the brain case. The keener hearing and better sense of smell demanded a bigger "computer" to handle increased data acquisition, which also required more energy—something the higher metabolic rate was able to supply. These early steps in the escalation of intelligence have set the mammals apart ever since.

Then came yet another critical step, initiated perhaps 100 million years ago, when one group of mammals developed a radically new device to act as interface between mother and developing fetus. One of the wonders of the animal world, the complex placenta (trophoblast) is a unique structure in which maternal and fetal circulatory systems transfer nutrients, oxygen, and waste products without inducing rejection of the fetus by the mother's immune system. (Because the fetus carries many different paternal genes it is, in essence, a foreign body.) Placental mammals include creatures as different as mice, monkeys, antelopes, bears, bats, and whales. The marsupial mammals (opossums, kangaroos, wallabies, koalas, and their kin) lack the complex placenta and have not been able to build so diverse a variety of body plans. Each little marsupial fetus must crawl out of its mother's womb before the mother's immune system is triggered to reject it. In contrast, having ample time within the womb, placental mammals can develop a wider array of body plans and build bigger brains.

Despite all these significant innovations, mammals remained at about the same level of relative brain volume for their first 150 million years. Except for the primates, it was not until about 50 million years ago that significant trends in brain enlargement became manifest in the fossil record of mammals. As in the birds, most mammals apparently were not subject to selection pressures for brain enlargement throughout the Mesozoic. After the demise of the dinosaurs 65 million years ago, mammals suddenly radiated into many daylight lifestyles. They grew bigger and more diverse, but it was only after this initial burst of diversification had leveled off that both prey and predator lineages of mammals began to demonstrate clear trends of greater encephalization.

Not all trends in vertebrate brain evolution have been toward bigger brains. Some lineages, such as the salamanders, have had a reduction in encephalization. Another notable example is the loss of the accessory olfactory formation and loss of lamination in the cochlear nuclei in dolphins, manatees, seals, bats, and primates (which includes us). These losses took place independently in the different lineages and reflect the fact that all of these animals abandoned their original terrestrial lifestyle. Whether living in the ocean, treetops, or caves, these mammals no longer used ground-based scent marking as a crucial part of their life activities and, consequently, the olfactory areas of their brains were no longer under selection to maintain these parts of the brain at earlier levels of organization. Similarly, many domesticated animals, such as cattle and chickens, have undergone a reduction in brain size, compared with their wild relatives. Dogs with heads as big as those of wolves have brains about 20 percent smaller than their wild and wily progenitors. These instances serve to remind us that selection is critical not only in developing new traits but also for maintaining them. The evolutionary escalation of intelligence had to be the result of complex interactions between environment, lifestyle, physiology, and, most important, continuing strong selection.

## BRAINS ARE EXPENSIVE

Bigger brains work best at higher temperatures, which is one reason why both birds and mammals maintain a constant warm body temperature, but those warmer bodies burn more fuel. You can forget to feed your pet turtle or snake for a few weeks, but forget to feed your canary or hamster for even a few days and you will have to bury it. In a fine review of animal intelligence, John Allman estimates that mammals require five to ten times as much food as a reptile of the same weight.[4] With sharply differentiated teeth, mammals cut, tear, and chew their food, helping fuel more rapid digestion. Many toothless modern birds keep their crop filled with pebbles to help grind up their chow. Both mammals and birds have larger intestinal surface area than reptiles so they can process food faster, and both have a four-chambered heart to distribute oxygen more efficiently. Just as significantly, both birds and mammals must support the early development of their young. Neither can allow their offspring to fend for themselves, as do most reptiles and amphibians. Large-brained offspring require large amounts of food to develop properly—food that they are unable to gather for themselves. Also, these rapidly growing young require high-quality food, with mammals providing milk rich in proteins, fat, and sugar, while most birds bring high-quality or regurgitated food to their young. Simply stated: building a bigger brain is expensive, requiring a richer diet and lots of tender loving care.

The cost of having a bigger brain is reflected in many aspects of an animal's life and lifestyle. By comparing small mammals that share the same body weight but have different brain weights, scientists have discovered important generalities. A rabbit may weigh the same as a small monkey, but its brain is much smaller and its life expectancy is much shorter. Because the small-brained rabbit is "cheaper-to-build" and requires relatively brief parental care, female rabbits produce more babies with each pregnancy and have litters more frequently than is the case with monkeys of equivalent weight. Not-so-bright rabbits keep ahead of extinction with a high reproductive rate, despite a short life span and high predation rate. Clever monkeys fend off extinction with longer individual life spans and a reduced predation rate, thanks to a bigger brain and, for most, a treetop habitat.

Much research has focused on the relation of brain weight to body weight in different-sized species. Comparing a large number of species has made it clear that the average rate of brain increase is three-fourths of the body weight increase. This power-scaling figure suggests that energy needs may be paramount, since metabolism also increases at a rate of three-fourths of the increase in body weight.[5] But things aren't really so simple. Mammals with longer gestation times and higher birth weights do have larger brains, on average, than similarly sized mammals with shorter gestation times and lower birth weights. Another important factor is diet. Leaf eaters tend to have smaller brains than fruit or insect eaters of the same weight. Thus, life history characteristics, food sources, and evolutionary history all play a role in determining a species' brain size. Among mammals, there are two important exceptions to the general patterns. Primates (monkeys and their relatives) in general, stand a bit above the mammal average while humans stand above both norms. This presents us with two questions: why are primates brainier, and what caused the human lineage to move even further along this trajectory of escalating brain size?

## PRIMATES ARE SMARTER

Early mammals had a brain-to-body ratio equivalent to today's opossums and hedgehogs, about four times as large as the reptilian average. For most mammals, this level of encephalization was maintained for over 100 million years, into early Cenozoic times. Fifty million years ago, the dog-sized ancestor of horses (*Hyracotherium*) had a brain-to-body ratio similar to that of modern opossums. As they became larger and more modern, the horse lineage reached an average modern mammalian brain-to-body ratio by about twenty million years ago. Luckily for us, there was one mammalian lineage that did not fit these general patterns.

From their beginning, the primates (apes, monkeys, lemurs, bush babies, and a few others) tended to have larger brains than other mammals, relative to their body sizes. This correlates with the fact that most mammals allocate about 5 percent of their metabolism to maintaining their brains while primates utilize 9 percent or more of their metabolism for the same function. A shift from a nocturnal lifestyle to the daytime pursuit of insects in treetops may have been the key initial step. Later, finding and remembering the location of fruiting trees would have reduced foraging time and promoted selection for a bigger brain. Jumping around among branches in the treetops demanded accurate three-dimensional vision with forward-pointing eyes. This was accomplished over millions of years by the shortening of the muzzle and shifting of the eyes into a position where their visual fields would overlap. A larger brain was helpful in processing two slightly incongruent images for accurate depth perception. Also, color vision provided a more realistic mental picture of a complex treetop world.

Because so much of their lives was spent among the slender tips of branches where trees produce their blossoms, primates evolved longer rear legs with grasping toes to hold on tight while reaching for yummy fruits with flexible forearms. Fortunately, and unlike many other mammal lineages, the earliest primates had retained all five digits on both their hands and feet (these are "primitive" features dating back to the amphibians). Separate digits with a flexible thumb and big toe were important in grasping slender twigs. Claws became transformed into more useful nails, while the grasping surfaces of hands and feet developed a ridged "friction skin" that facilitated holding onto smooth branchlets. At the same time, arms and shoulders were developing new kinds of flexibility; and a versatile wrist with nimble fingers allowed for better grasping as well as the careful examination of fruits or bugs. Hand-to-mouth feeding, after careful examination, is characteristic of all primates. Later, with the evolution of apes, came a flatter chest and broader shoulders, permitting the arms to rotate in a wide circle. Finally, with a few further refinements, one species learned how to walk on its hind legs and throw rocks with the front ones—but we're getting ahead of ourselves.

None of this monkey business would have been possible without a significant addition to the landscape: flowering plants. Neither tree ferns, cycads, ginkgos, nor conifers could provide such an abundance or variety of nutritious nuts, juicy fruits, and nectar-filled flowers as did the flowering plants. These were the features that attracted so many insects into the treetops, soon to be followed by the earliest insectivorous primates. Robert Sussman argues that it was the later development of larger fruits by flowering trees, as well as a more omnivorous diet, that played a key role in the evolution of those features so distinctive of monkeys and their kin. Our stereoscopic color vision has more to do with finding and carefully examining fruits and nuts than catching bugs, claims Sussman.[6]

Prolonged nurturing of the young by females and complex social interactions also played important roles in having early primates reach higher rates of encephalization. With forward-pointing eyes, primates reduced their lateral vision. Thus it became important to travel in small groups where extra pairs of eyes could survey all directions for danger. As the primates continued evolving, there were two instances when encephalization took additional steps forward. The first increase took place around 20 million years ago, in the fossil group called *Proconsul* and its immediate predecessors. These animals, probably originating in Asia and living in moist evergreen forests, were among the earliest of the apes. Not only did *Proconsul* have a slightly larger brain, it lacked a tail, as do all the living hominoids (gibbons, orangutans, gorillas, chimpanzees, and ourselves). The legs were strengthened to support a more upright posture, and the back became stronger. Early apes probably spent more time swinging from branch to branch (called "brachiating") in wide-crowned tropical trees. If you haven't seen gibbons actively swinging through the branches, whether in a zoo or in nature, you have missed something very special. And here is where the flowering plants made another significant impact on the evolution of our ancestors. Not only do conifers lack flowers and a similar variety of tasty fruits, few nonflowering trees have strong, broadly spreading branches allowing a larger animal to swing freely from branch to branch, an activity that gave our particular ancestors wide shoulders, long arms, flexible elbows, and long dextrous fingers. Flowering trees not only fostered the origin of primates, but I would suggest they also made possible the later development of our more upright swinging ancestors.

## WALKING ON TWO LEGS

After *Proconsul* and the origin of apes, the next critical event for our lineage was spending more time on the ground. Our nearest relatives, chimps and gorillas, spend a fair amount of time on the ground, and both have developed a four-limbed "knuckle-walking" mode of travel when they're on the ground. Our ancestors may have walked in much the same way, but they soon became more upright and two-legged, probably sometime around 5 or 6 million years ago. The earliest fragmentary fossils are about 5.5 million years old, but a somewhat more recent stage is much better documented thanks to the fossil called "Lucy," given the scientific name *Australopithecus afarensis*.[7] The genus was originally described from a fossil cranium found in 1924 in southern Africa, hence the Latin name *Australopithecus* (southern ape). Discovered fifty years later, Lucy lived about 3.3 million years ago in Ethiopia's Afar Depression, part of the Rift Valley south of the Red Sea. Today, the dominant vegetation of the Afar depression and lower Awash Valley is subdesert grasslands and thornbush, but fossil pollen indicates that the climate was very different in Lucy's time, when both moist evergreen forest and open wooded savannas flourished.

Most human fossil discoveries comprise a few bits and pieces, but Lucy is extraordinary in having had 40 percent of her skeleton recovered.

An example of "mosaic evolution," her skeleton reveals a combination of both primitive apelike characteristics and more humanlike features. While Lucy's face, teeth, and cranium were very much those of an ape, her pelvis and backbone clearly indicate that she walked upright. The lateral extension at the top of her thigh bone is shaped like ours, and quite different from that of chimps and gorillas. Another exciting discovery, not far from where Lucy's bones were found, revealed the scattered fragments of about a dozen individuals of her species that may have died together in a flash flood. Taken together, these Ethiopian fossils have provided us with a very good picture of this early stage in human evolution. Their feet resembled ours, but they still had traits of a more apelike ancestry. Their toes, for example, make up about 30 percent of the length of the foot, a smaller percentage than chimps (36 percent) but considerably larger than modern humans (23 percent). Long curved fingers and strong shoulders make it clear that these creatures remained adept at climbing trees, probably to escape danger and for spending the night. Though perhaps incapable of making stone tools, Lucy's hand structure should have been quite effective in throwing stones and manipulating sticks.[8]

All told, these remains portray an upright species having a very apelike skull with a brain volume of about 400 cc, not very different from that of a chimpanzee. (Brain volume in chimps ranges from about 380 to 440 cc. When describing brain volumes we'll use cubic centimeters or cc; a cubic inch holds 6.4 cubic centimeters.) The significance of Lucy and her associates is that they provide clear evidence for an upright stance with significant changes in pelvis, spine, and feet *long before* major changes had taken place in brain size. The extraordinary fossil footprints found at Laetoli in Tanzania represent the same evolutionary stage a few hundred thousand years earlier. These footprints clearly reveal a humanlike stride and that the arch of the foot was well developed, a key element in producing a cushioned impact and resilient forward step. Besides our enlarged brain case and unusual pelvis, it is the anatomy of the foot that distinguishes humans most clearly from all other living primates. Lucy and her Australopithecine kin represented a very successful stage in our lineage. They persisted on the east African savanna for well over two million years, and ranged from southernmost Africa to its northeastern corner in Ethiopia and Eritrea.[9]

The differences between Lucy (*Australopithecus afarensis*) and the earlier discovered but slightly more modern fossil material of *A. africanus* (based on the "Taung child" cranium from South Africa) are subtle. Fossils assigned to *Australopithecus africanus* differ from *A. afarensis* in having a slightly larger (460 vs. 400 cc) brain volume, smaller incisors, a first lower premolar with two cusps (instead of the apelike single cusp), and the absence of gap in front of the upper canine teeth. This gap in apelike jaws leaves room for the large lower canine when the mouth is closed. It seems clear that the more modern *A. africanus* was becoming less apelike with time. But let's look at a more fundamental question: what might have impelled Lucy's ancestors to become upright and two-legged in the first place?

Our closest relatives, the chimps and gorillas, live in evergreen forests in central and western Africa. It's likely that their ancestors were also confined to heavily forested areas. Yves Coppens has argued that geological rifting and a transition to a drier climate, beginning about 6 million years ago, was the determining factor separating our hominid lineage from that of the chimp and gorilla.[10] A minor aspect of the Earth's system of dynamic plate movements, rifting and its associated faults played a major role in the creation of elevated highlands along much of eastern Africa. Almost 6,000 miles long, the rift system ranges from Israel and Jordan, southward along the Red Sea, through Ethiopia and eastern Africa to Mozambique. As the rift highlands gained in elevation, they impeded the movement of moist equatorial winds and contributed to a drier climate in eastern Africa.

With gradual drying, lowland areas became dominated by thornbush, open acacia woodlands, savannas, and subdesert grasslands, though evergreen forest often persisted along river margins. Reduced rainfall inevitably reduced plant productivity. Fruits, nuts, and tubers became more widely scattered and difficult to find; animal numbers decreased. This was a time, five to six million years ago, when major changes are recorded in African fossil faunas, with some species going extinct, new immigrants entering the region, and local speciation in those groups already adapted to drier vegetation. It was under such conditions that the surviving African apes had split into two branches. One group continued to amble through evergreen forests walking on its knuckles, while the other got up on its hind legs and wandered out into the savanna. As resources became more difficult to gather, ground-living apes in dry habitats had few choices. Reducing group size might be one strategy in the face of fewer resources, but this would make the group more vulnerable to predators. Getting up on one's hind feet reduces the energy needed to cover longer distances, allowing the bipeds to retain a larger group size. Anthropologists Lynne A. Isbell and Truman P. Young have argued that these effects and ongoing competition between groups played a critical role in the origin of bipedalism.[11] Moving on two legs allowed this new lineage to move into drier and more open vegetation—to survive where no ape had lived before.

Moving on two feet had a number of additional advantages. An upright posture reduces the heat load under a high tropical sun. Having a relatively large brain, sensitive to overheating, the human lineage needed to develop a more complex network of blood vessels to cool the brain. Cooled by evaporation from face and head, blood circulation would now moderate brain temperatures.[12] The importance of cooling is also reflected in the fact that humans have lost most of their body hair while having acquired a density of sweat glands greater than any other mammal. Unfortunately, sweating means losing a lot of water. Despite the advantage of bipedalism for dry open habitats, stringent water requirements meant that our progenitors could not stray far from permanent water sources.

Standing upright also exposed us to cooler breezes and provided a higher vantage point for looking over grass and low shrubbery. Turning on two feet makes surveying one's surroundings much quicker, facilitating the continuous monitoring of broad vistas. To accomplish this, our eyelids developed a more elliptic opening for horizontal scanning, quite different from most other primates. Together, the many advantages of bipedalism would help a larger troop sustain itself over an area of few and scattered resources. In turn, such a group would be under strong selection for a better memory of local terrain, stronger social bonds, the sharing of food, improved communication, and united action in hunting and defense. Becoming upright, the human spine developed a curvature and the chest was flattened, improving the distribution of body weight. The pelvis broadened for the attachment of powerful thigh muscles and to support the intestines. The articulation between skull and terminal vertebra shifted to a position underneath the skull to better balance the head. The legs became longer and the foot changed profoundly: bringing the big toe alongside the central axis, reducing the little toe, developing an arch, and extending the heel bone backward for stronger muscle attachment.

Most important for humans, walking on two feet emancipated the hands and set the stage for a quantum leap in the escalation of intelligence. With the forelimbs free, the arms could be used to carry food back to the group. Hands and fingers were able to fabricate and manipulate tools more easily. It was now possible to challenge predators with a barrage of stones or with raised clubs. Perhaps even more significantly, gesticulating arms, hands, and fingers afforded new possibilities for visual communication—and paved the way for speech.

## BUILDING A BIGGER BRAIN

Though the fossils are fragmentary, Lucy's stage and that of her immediate predecessors was a successful species style for at least a million years. Lucy's lineage then gave rise to two new and different lineages (see figure 2 on page 185). One derivative line, the "robust" australopithecines (*A. robustus* of southern Africa and *A. boisei* of eastern Africa), developed much more massive skulls with huge jaws and very large molars, probably as a result of adapting to a vegetarian diet of hard-to-chew plant foods. The top of the skull had a longitudinal crest to help anchor powerful jaw muscles. Other aspects of the skeleton were very similar in size and structure to our own lineage. Restricted to Africa, the robust australopithecines (sometimes placed in the genus *Paranthropus*) flourished for more than a million years. They exhibited only slight expansion in brain volume as they developed larger bodies, before becoming extinct about 1.4 million years ago. Our own lineage retained a smaller jaw and teeth, indicating that we continued to eat a varied diet. It didn't take this particular lineage of two-legged omnivores very long to find that the many large savanna herbivores were an excellent source of highly nutritious food. Beginning most likely as scavengers, our lineage would remain omnivorous and soon become efficient hunters. In contrast to the big-jawed robust australopithecines, our lineage began to develop bigger brains; the fossil representatives of this lineage are placed in our own genus, *Homo*. (The two upright genera, *Australopithecus* and *Homo*, are called hominids.)

Surprisingly, the brains of humans develop at *exactly the same rate* as do the brains of chimpanzees, gorillas, and orangutans. The significant difference is that human infants are born quite underdeveloped with their skull sutures open, and their brain continues growing rapidly during the first year of life. Thus, our adult brain is about four times larger than when we were born, whereas the brains of mature primates are only twice the size of a newborn's. That's the good news. The bad news is a completely helpless baby; mothers cannot cradle or carry the baby *and* climb a tree. Paleontologist Steven Stanley has suggested that until the troop could defend itself on the ground, it was not possible for a mother to raise a helpless infant—an infant unable to cling to her back as she scampered up a tree to avoid a predator.[13] The fact that Lucy and her bipedal predecessors roamed the African landscape for more than two million years, and that her bone structure indicates she was still an efficient climber highlights the importance of Stanley's suggestion.

Bipedalism itself may have played a role in these changes. As the foot became more adapted to walking, the big toe was reduced in size and closely aligned with the other toes. No longer able to grasp with their feet, little humans could not hold on to their mother's hairy back in the same way a baby chimp could, using both hands and feet. Here was another reason why human mothers had to hold their babies, and it provided a strong selective pressure to develop better security on the ground. Only after the stick-waving troop could defend itself on the ground and through the night could the final step in escalating intelligence proceed. Defense during the long African night may have involved sleeping on cliffsides or constructing encampments enclosed by a barrier of thorny branches. Not until this level of group defense had been achieved could our ancestors give birth to helpless infants whose brains would continue to grow, develop, and learn.

## BRAIN SIZE AND INTELLIGENCE

Intelligence has always been a slippery concept and we'll use the word in a broad sense here. Harry Jerison, who has studied brain evolution intensively, calls intelligence "the capacity to construct a perceptual world," and this must certainly be the bottom line for all brains, large and small.[14]

Let's consider the activities of a sand wasp. She excavates a burrow in the dunes, covers the burrow entrance, flies off to hunt for prey, returns after about twenty minutes carrying her victim, circles the area to relocate her burrow, sets down her prey, uncovers the entrance, drags the victim within, and lays an egg on the paralyzed prey. This is no small achievement, especially when one considers that this creature never went to wasp school. Her ability to find an appropriate site, dig a burrow, hunt for prey, and find the burrow again requires many accurate "pictures" of the world in which she lives. Remarkably, her burrow is often made in broad dunes covering many acres. What's also mind-boggling about this life-in-miniature is that the wasp's onboard computer is about the size of the head of a pin. More baffling yet, this entire system is transferred from generation to generation by information borne on genes neatly arranged along her chromosomes.

Larger brains are even more amazing. On a very dark and rainy night in Costa Rica, our flashlights picked up what appeared to be birds flying just above a broad jungle river. I couldn't understand how anything could fly so close to the river's surface under such very dark conditions, until a guide pointed out that we were watching fishing bats. They fly perfectly level just above the water's surface in the dark of night. Using their high-pitched calls for echolocation, they can detect ripples on the surface, catching minnows with their clawed hind feet. Recounting my amazement at seeing these creatures in action, a bat specialist responded that what impressed him most was being deep in a cave with thousands of bats flying about in absolutely lightless space without colliding. Not only was each bat forming an accurate "image" of its surroundings using the reflections from its high-pitched calls, but each bat was able to separate the returning echoes of its own signal from the signals and echoes of all the other bats. The high-speed analysis of returning echoes by a flying bat is an awesome computational feat, thanks to a "computer" no bigger than a pea.

When it comes to our own brain, we use it so effortlessly that we fail to appreciate its complexity. Our vision appears to be simple and straightforward, but experimental studies of vision and brain function have shown it to be an amazingly complex process. This information-processing protocol begins within the eye itself where the signals from 100 million rods and cones must be collated and processed before being transmitted by an optic nerve having "only" a million nerve fibers. Years of laboratory experimentation with domestic cats and small monkeys have shown how a number of different batteries of neurons and different areas of the brain contribute to the creation of what appears as a single comprehensive moving image. Unusual movements in the environment are noticed immediately, even when they are at the edge of the peripheral field. The fovea, a small central area where we see sharply, scans the moving element repeatedly as the brain analyzes the object. Failure to recognize a predator could be fatal, but overreacting to gusts of wind wastes precious energy. Decisions need to be made quickly and economically. Having to distinguish the intermittent cadences of a gusting wind from the interrupted rhythm of a quietly stalking predator may have created in our mind those faculties we now use to make music and explore the deep structure of mathematics.

Coupling the data streams of sound and sight for quick evaluation and response is essential for the survival of most animals. Because we do this continuously and without effort, it appears to be a unitary process, masking the complexity of many interactive neural networks. Different areas of the brain and visual apparatus must work in real-time synchrony, and many other aspects of body and mind contribute as well. When you suddenly change the position of your head from vertical to horizontal, your field of view does not flip 90 degrees, thanks to mental processes keeping you in tune with the outside world as it really is. Likewise, when someone unexpectedly shuts off the light and you find yourself in complete darkness, you still retain a mental picture of your surroundings and can move accordingly. Modern science still cannot explain how it is that we understand what we see.[15]

The minuscule brain of the sand wasp and the little brain of the bat highlight another problem: brain size is not a very accurate measure of intelligence. Big people with bigger brains are not smarter than little people with smaller brains. But the fossil record provides us with only limited information; the behavior of extinct mammals cannot be tested. Brain size and hints of its organization inferred from endocasts of fossil skulls are our only clues of escalating mental abilities over millions of years. Fortunately, bigger brains are generally correlated with more flexible behavior and more complex lifestyles when making comparisons across species. Increasing brain volume in fossil hominids over the last three million years provides critical data in understanding the last advance in the escalation of intelligence. Compared to our primate relatives, and taking our body size into account, we humans have a brain about three times bigger than it ought to be. We are the only species on the planet that has made the behavioral leap to an abstract system of verbal utterances able to describe and reconstruct a site or event so that its salient features are communicated to someone who has not shared that experience. How did such an amazing ability arise?

## HOMINOID INTELLIGENCE

Vervet monkeys have three alarm cries that distinguish between their most dangerous predators: eagles, snakes, and leopards. Each specific cry will elicit an appropriate response in other members of the troop. This clearly demonstrates the vervet monkey's perceptual acuity, but it is not clear whether the response is due to inherited behavior or learned associations. Humans can do something far more significant: they can come back to camp and tell the rest of the group exactly what lies beyond the nearby hills. They can communicate their experience in rich detail to their comrades, and the troop can then decide whether or not to move to the distant valley.

Human abilities did not spring full blown from an average mammalian brain. Old World monkeys are smart; they are known to deceive each other though they may not really understand what they are doing. The more primitive New World monkeys do not display deceptive behaviors. The great apes often use deception and counterdeception, taking into account the other ape's perspective. The ability to infer what others are thinking is an important aspect of intelligence in chimps and ourselves; it is a skill autistic children lack. The ability to "mind read" the attitudes or intentions of others probably developed as social interactions became an important aspect of creating an effective and cooperative troop. Wolves have developed similar skills, with the result that their descendants, domestic dogs, are often keenly aware of their master's every mood.

Nearly all of our field and laboratory research with chimpanzees is focused on the common chimpanzee (*Pan troglodytes*), which ranges in disjunct populations from western Africa to the western edges of the East African rift system. Common chimpanzees live in groups of variable size dominated by males. The males can be very aggressive, both within and between groups, and they are consistently larger than the females. In contrast, male bonobos (*Pan paniscus*), do not display such aggressive behavior, do not play so dominant a role in the troop, and are quite similar to females in facial features. In addition, bonobos, who are about the same size as the more common and widespread chimpanzee, are more arboreal, more shy, and have a very restricted geographic range in central Africa, south of the Congo River. For these reasons, bonobos have been the subject of only a few field and laboratory studies.[16]

Studies of common chimpanzees in the wild have shown troops in different areas and environments to have very different ways of doing things (sometimes called "traditions"). Tool use has been documented in many groups. A western group of chimps has been seen carrying small rocks up into the trees, where the rock is used as a hammer to crack open hard nuts against large branches. Another troop has been observed using two tools: one to crack open oil-palm nuts, and another to pound stems to gain access to the pith. Also, wild chimpanzees have been seen carefully folding and eating leaves of a nonfood shrub (*Vernonia amygdalina*), the same leaves some African cultures use as a medicine to treat intestinal parasites. In laboratory environments, carefully trained common chimpanzees have demonstrated impressive problem-solving and communicating skills. Most of these skills have never been seen in wild populations, but there is evidence for creative teaching and careful imitation by chimps in the wild. The significance of research on common chimpanzees trained in laboratory and domestic human settings is that we have come to see far more sophisticated abilities than have been seen in field observations, and thereby gained greater appreciation of their mental capacities.

Not only do chimps adhere to "rules of conduct" in their own affairs, but they quickly learn rules imposed by humans—and remember when those rules are broken. Frans de Waal reports of an instance where the "rules" were broken and the rule breakers were punished. The human rule was that dinner feeding would be delayed until *all* the chimps had come in from outdoors. In this case two adolescent females stayed out, making the others wait two hours for dinner. These youngsters were kept apart from the main group during that night, but when reunited outdoors the next day they received a beating from the entire group, which had seemingly remembered the long wait. "Needless to say" remarks de Waal, "they were the first to come in that evening"—a nice example of the efficacy of corporal punishment for those who've broken the rules.[17]

A more uplifting example of ape behavior was recently reported in newspapers and TV after a three-year-old boy fell into the gorilla enclosure at Chicago's Brookfield Zoo. Having fallen more than twelve feet, the child lay prostrate and unconscious on the concrete floor. Binti Jua, a six-year-old lowland gorilla, moved quickly to the child and cradled it in her arms. Perhaps fearful that she might upset other gorillas in the same enclosure, Binti Jua oriented herself facing away from them and then carried the child to a door where keepers were able to retrieve the still-unconscious boy. The entire sequence, taking but a few minutes, was made more poignant by the fact that Binty Jua was carrying her own child on her back.[18]

Laboratory work and intensive training of chimpanzees indicate that they may possess the ability to understand both tangible and abstract concepts. Primatologist Gordon Gallup carried on a series of experiments in which vervet monkeys and chimps became familiar with their own image in mirrors. He then marked their foreheads with paint, without the animals realizing what had been done. Upon seeing themselves in the mirror, the chimps immediately touched the painted spot on their foreheads but the vervet monkeys never responded to the change in their image. Self-awareness is clearly part of a chimp's intellectual capacity. A sequence in the TV nature film *Monkey in the Mirror* shows a young chimp's first encounter with a large mirror. At first she treats the image as another chimp but finds that it is a flat surface with no other chimp behind it. She is confused by the synchronous actions of the image and herself; slowly, as she touches the mirror, she appears to realize it is a reflection of herself. This discovery is followed by a wonderful sequence of looking at herself eye-to-eye, followed by further intense self-examination, beginning with face and mouth and ending with a careful review of those difficult-to-see areas at her bottom end.[19]

What appears to be self-recognition in chimps indicates that human self-consciousness is not a distinction that came magically into being from a previous void, as some authors would have us believe. Our brains and our consciousness have been honed by natural selection to make us effective members of complex social groups and help us deal with an often hostile environment. The "mind-brain problem" frequently discussed by philosophers and psychologists is a problem only for those who refuse to consider an evolutionary explanation. The evolution of intelligence began when a stretch of DNA managed to code for the production of proteins that would, one way or another, help the earliest bacterium move away from environmental threat or toward something to eat. The nervous watchfulness of an antelope or the silent stalking of a lioness are highly evolved expressions of animal intelligence and clearly indicate a sophisticated level of awareness. With a deep self-awareness, a phonetic system capable of communicating both specific and abstract concepts, and the ability to imagine the future, humans have taken consciousness to its highest level.

## BRAIN AND LANGUAGE

Having studied comparative brain evolution in vertebrates, Jerison concluded that the primary function of language, as well as the enormous development of the neocortical region of the human brain, has been the communication of perceptual knowledge.[20] With language we can describe not only the presence of prospective prey but also their movements, the nature of the vegetation they are moving through, and a strategy for intercepting them. Our brain not only pictures reality, it can conceptualize and communicate that picture. Indeed, spoken language may have been "the supreme human technology"—the primary adaptive breakthrough from which all our other achievements arose.[21]

Neurological studies indicate that our ability to conceptualize objects began with visual recognition, and the visual centers of the brain still play an important role in conceptual brain activity. Developing vocal signals and a grammar to communicate more complex and abstract concepts required a bigger machine with more wiring. Recent studies using echo-planar functional magnetic resonance imaging show that more difficult conceptual problems involve a wider range of brain areas. Experimental subjects were given sentences of increasing linguistic complexity, and the imaging clearly showed the amount of brain activity increasing with the difficulty of comprehending the sentences—a graphic demonstration of the reality of having to "think harder."

Robin Dunbar argues that language played a key social role in humans, replacing the practice of intensive grooming in other primates. Grooming has the important function of maintaining status relationships, affording a means of reconciliation after a squabble, and developing trust. Language provides us with a less time-consuming "vocal grooming" to facilitate social cohesion, argues Dunbar.[22]

Vocal communication requires the ability to form a variety of sounds, especially vowels, allowing us to divide the ongoing sequence of sounds into separate syllables. A major anatomical change in the human lineage has been the lowering of the pharynx, producing a longer chamber, the larynx. The larynx allows a larger volume of air to be shaped by mouth, tongue, and lips into a great variety of sounds. That's the positive side of the ledger; the negative side is that having a lowered pharynx can be awkward and even deadly. Unlike chimps and other mammals, adult humans cannot breathe and swallow at the same time. The enlarged larynx and an upright posture make it much easier for food to become lodged in the windpipe. In the United States each year, over a thousand people die because of food caught in their throat. The Heimlich maneuver, to free food caught in the windpipe, is a peculiar human need; it is the price we have paid for articulated speech.

Human speech is based on coordinated use of the brain's motor cortex areas controlling mouth, tongue, and larynx. It is highly unlikely that such a sophisticated neurological-anatomical skill could have sprung into being in a single fortuitous event. Gradual descent of the pharynx during our youthful years and children's slow acquisition and mastery of language may reflect a long and gradual evolutionary development. The one-year-old speaks three to five words, the two-year-old about fifty, the three-year-old about a thousand; by five, grammar and comprehension are well developed.[23]

Many years of research on brain functions suggest that language evolved from an existing object-recognition system that originated in our most remote mammalian ancestors. Such an object-recognition system was the basis of our ability to characterize and classify real visual objects at first, and more abstract concepts later. Communication by gestures of the body, face, and hands provided the neuromuscular patterns from which spoken language would emerge.[24] Modern neurological studies have shown that visual pathways, various memory areas, premotor regions, physical and semantic attribute areas, and the limbic system all contribute to language function. Positron emission tomography has shown that congenitally deaf people use the same areas of the brain when "hearing" sign language as we do when we hear spoken language. Merlin Donald provides a detailed and well-developed scenario for the evolution of human cognition and speech in his book *Origins of the Modern Mind*.[25] Unfortunately, he divides the evolution of increased encephalization and symbolic language into three clearly separate stages, implying sudden innovations with each new level of ability. Donald correlates these separate stages with the rather arbitrary taxonomic constructs: *Homo habilis—Homo erectus*—modern *Homo sapiens*. Such a view seems a little too pat. According to Donald, Philip Lieberman, Ian Tattersall, and others, people at the *erectus* stage lacked the linguistic abilities that suddenly burst into flower with early *Homo sapiens*.[26] None of these authors explains why such extraordinary new abilities as symbolic thought and speech should suddenly have burst into being with modern *sapiens*-stage people.

Careful analysis of brain casts made from the limited number of cranial remains of *Homo habilis*, the earliest member of our genus, indicate an enlargement of a lateral part of the frontal lobe and the parietal area along the left side of the cerebrum.[27] These are the same regions of the modern human brain called "Broca's area" (behind the left eye) and "Wernicke's area" (above the left ear). Accidents involving these areas can result in loss of language functions. Close to the brain centers controlling lips, tongue, and vocal cords, Broca's area, if damaged, results in severe speech impairment. Farther back, close to the region of the brain involved in hearing, Wernicke's area, if damaged, results in fluent but incoherent and meaningless speech, and in the inability to understand speech. Recently, the hypoglossal canal at the base of the skull, which carries nerves to the tongue, has been shown to be very similar in size to our own in late *erectus*-stage fossils over 500,000 years old, and much larger than in chimps, gorillas, and two-million-year-old hominid fossils. Such data strongly support Jerison's claim that the development of human speech was closely correlated with the evolutionary enlargement of the brain over the last two million years.[28]

The notion that language is a recent acquisition of *sapiens*-stage people is a reflection of the popularity of the "punctuated equilibrium" view of evolution, where changes are said to come in sudden spurts between long periods of stasis. A number of authors believe that language arose during the last "punctuation" in our history—the emergence of modern *Homo sapiens*. But confining language to the latest, and quite arbitrary, taxonomic stage called *Homo sapiens* seems naive. There are too many subtle interactions between vocal cords, larynx, facial muscles, tongue, hearing, and brain to expect anything but a slow and gradual refinement of pre-existing abilities. Suggestive evidence for a gradual increase in human cognitive and social abilities comes from an unusual recent discovery.

Ancient wooden spears, found deep in a German mine, give us new insights into life long ago.[29] Because plant materials are broken down rapidly under normal conditions, wooden artifacts are rarely preserved in the archaeological record. An extraordinary find, these six-foot-long spears were preserved in water-logged deposits. Carefully crafted with center of gravity a third of the length behind the point, just as in a modern javelin, these devices were used by the hunters of large game. Interestingly, the apex of the spears was carved from the *base* of young spruce trees, where the wood is especially dense and hard. The skill and thought exhibited by those who carved these tools is especially significant when one considers that they were found in deposits dated between 400,000 and 380,000 years ago. A major discovery had they been only 40,000 years old, the spears' true age represents a unique glimpse into a much more ancient time. These were "late *erectus*" people hunting big game during a brief warmer interglacial period in northern Europe. The fashioning of such a sophisticated weapon by an ancestor of Neanderthals is not likely to have happened in a single moment of creative brilliance. Rather, the best place to position the center of gravity, and the best way to carve a young spruce tree, were probably learned over time and *communicated* by utterances to finally become speech. Such a discovery is consistent with a gradual and coordinated evolution of increasing brain size among our ancestors over the last million years.

"Grammar is a protocol," declares Steven Pinker in his book *The Language Instinct*, ". . . a protocol that has to interconnect the ear, the mouth, and the mind, three very different kinds of machines. It cannot be tailored to any one of them but must have an abstract logic of its own."[30] He prefers to use "the admittedly quaint word instinct" to describe the innate ability of children to learn language, and he points out that it conveys the notion that people know how to talk in much the same way as spiders know how to spin their webs. This makes a lot of sense. Human babies begin babbling a few months after birth; they do this instinctively, their continued vocalizations and mimicry of the language they hear are all part of an internally driven program. Parents do not teach their children how to speak. These inborn behavior patterns are part of our genetic heritage—just as is our large brain, along with all the other features that make us modern human beings. Pinker's point of view seems eminently reasonable: our ancestors refined their unique linquistic skills slowly, over thousands of generations. But what, exactly, was the shape of this trajectory?

## HUMAN BRAIN EXPANSION: GRADUAL OR STEPWISE?

Three and a half million years ago, Lucy had a cranial capacity of about 400 cc. One million years later, *Homo habilis* had an average brain size of between 500 and 600 cc. At about 1.6 million years ago, early *Homo erectus* had a brain averaging 900 cc. *Today, humans average between 1,300 and 1,400 cc, a more than threefold increase in 3 million years.* Since Lucy and her comrades were quite small, some of the early increase in brain size was simply due to our ancestors becoming larger. Although our ancestors reached the general size of modern people around 2 million years ago, the fossil evidence indicates that the human brain has more than doubled in size since then. Humans, and humans alone, reached a level of intellectual prowess not matched by any other species on this planet. Today an average twenty-year-old American is estimated to have a vocabulary of 45,000 words, running on a computer with 100,000 million neurons linked by who-knows-how-many interconnections. This was no ordinary evolutionary achievement; but was it gradual or stepwise?

In his monograph on *Primate Origin and Evolution*, Robert Martin concluded that the "increase in brain size has been a fairly continuous phenomenon throughout hominid evolution, and present evidence does not indicate any sharp change in tempo at any given point."[31] However, it is important to remember that the fragmentary fossil record allows for a variety of other interpretations. The notion of continual gradual increase in brain size over the last three million years is challenged by John Kappelman.[32] Measuring the orbital aperture (eye socket) to estimate the body weight of fossil hominids, he concludes that there were three steplike increases in brain size to body weight relationships. The first was at Lucy's stage, the second with the origin of the genus *Homo*, and the third after the rise of more modern *Homo sapiens* people about 100,000 years ago. In fact, Kappelman insists that the last stage of brain increase was entirely due to a reduction of body mass, with a concomitant "increase" in relative brain volume.[33]

A very different approach to the question of increased cranial expan-

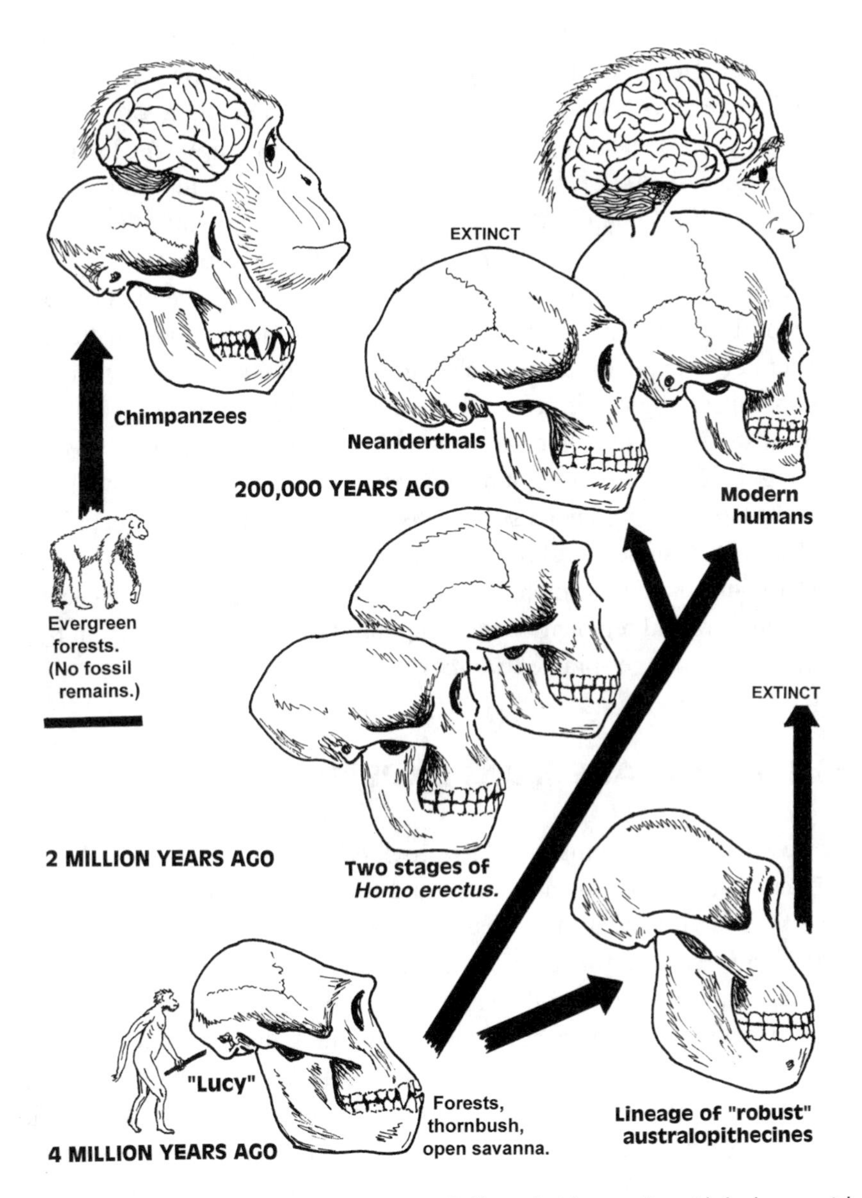


**Figure 2.** Hominid cranial evolution. "Lucy" (*Australopithecus afarensis*) had a cranial volume of about 400 cc, similar to modern chimpanzees. Average cranial volume expanded from about 800 cc in early *Homo erectus* to as much as 1,500 cc in Neanderthals, and 1,300 to 1,400 cc in modern humans.

sion over the last 1.6 million years is taken by paleoanthropologist Grover S. Krantz.[34] He prefers to lump all fossils in the 1.5 million years prior to 40,000 years ago into *Homo erectus*, and he divides the fossil material into four geographical regions: Africa, Europe-western Asia, China, and India-Southeast Asia. He then separates the fossils into three time periods: early (1.6 million to 800,000 years ago), middle (800,000 to 300,000 years ago), and late (300,000 to 40,000 years ago). This analysis, although simplified, is instructive. From a common average of about 900 cc 1.6 million years ago, data from all geographic areas show a pattern of accelerated cranial expansion. The important point is that, while the regions differ slightly in average cranial capacity at any point in time, they all show increases in brain volume in a similar accelerating way. By treating the geographical regions separately, his data clearly demonstrate that *Homo erectus* was not a static phase of evolution, and that there is no evidence for a sudden worldwide increase in brain size with the advent of modern humans. Furthermore, the continuity in the regional curves seems to support gradual regional trends as opposed to sudden replacement by a more modern, bigger-brained *sapiens*.

## BIG BRAINS ARE A BIG EXPENSE

The escalation of intelligence has provided humans with the ability to communicate in sophisticated ways and forge new solutions for dealing with their environment, but it has also been very costly. Not only is the large brain extremely expensive to construct, requiring a rich diet for expectant and lactating mothers, the brain's energy demands are continuous. Though only 2 percent of our body weight, the brain burns 20 percent of the energy required to sustain us. When we are deprived of oxygen for even a short period, the brain is the first organ to suffer irreversible damage. Neurons need to "stay charged" to be functional, consuming energy even when inactive. We do much of our thinking with the reddish purple outer cortex of our brain; it becomes "gray matter" only in death, when drained of the nourishing flow of blood that continuously sustains it.

The large head of the human infant makes birthing difficult and dangerous; childbirth has been a major killer of women throughout history. On average, a 200-pound female gorilla gives birth to a 4-pound baby, while human females weighing 90 to 150 pounds give birth to babies weighing 6 to 9 pounds. The probability of death associated with childbirth during a woman's lifetime is between 1 in 25 and 1 in 50 in Third World countries today.[35] In these countries, death in childbirth makes up about 25 percent of all deaths among women of childbearing age. The death rate from childbirth among wandering hunter-gatherers must have been even higher. "The ninety-six hours or so following the onset of parturition [birthing] constitute the greatest single period of mortality risk that the typical human will ever face," writes Peter Ellison in his review of human reproductive biology.[36] With difficult birthing, caring for helpless infants, and acting as primary caregivers over a long period of nurturing, women bear much of the "cost" of our having the world's most complex organ between our ears.

As noted, building such big brains requires a high-quality protein- and fat-rich diet over many years. Humans have distinctly shorter intestines than do chimps, gorillas, or orangs. The reason is simple: the apes eat a higher percentage of leaves. This is lower-quality food requiring more time to be digested. Humans, who eat richer food, do not need so long an intestine, but we *do need* rich food. During the first year of our life, it's been estimated, half of our energy requirements are devoted to nurturing the growing brain. The primary task for human hunters was to garner as much animal fat and protein as they could get their hands on. Gathering nuts and chewing on tubers would not suffice.[37]

Many modern anthropologists, perhaps inspired by the desire to deny a more violent human past, claim that it was not until modern *sapiens* people came along that we humans became true hunters. This consensus posits that earlier people had merely scavenged the kills left by larger carnivores. But such a view ignores the importance of animal protein and fat in human development, fails to take into account the scarcity of abandoned large-carnivore kills, and minimizes the difficulty of satisfying our protein requirements with a preagricultural plant diet. Microanalytic studies of discarded bone remnants, both those left by Neanderthals 50,000 years ago and those eaten by *Homo erectus* one million years ago, indicate that a majority of the killed animals had *not* been chewed on by large carnivores. Even stronger evidence, reported in January 1999 and using carbon isotope ratios in the teeth of *Australopithecus africanus*, claims that their diet was rich in animals that ate open-savanna "$C_4$" grasses. Had these early hominids been eating mostly fruits, nuts, and tubers, their teeth would have had a larger "$C_3$" signature.[38] Beginning more than two million years ago, building a bigger brain required a richer diet; it seems likely that we humans have been active hunters ever since such early times.

Poor diet, especially protein deficiencies, can have devastating effects on the intellectual development of children. However, given a good diet and a nurturing environment, how much does heredity influence human brain power? A Swedish study has recently reconfirmed the role for a strong genetic component in our cognitive abilities. This study compared 110 pairs of identical twins with 130 pairs of nonidentical same-sex twins. What made this study different was that none of the subjects had significant mental impairment, and all the pairs of twins were more than eighty years old.[39] Results of this analysis, consistent with many similar earlier studies, found that slightly more than half of the variance in cognitive ability could be accounted for by heredity. A shared environment accounted for about a third of the variance, while the remaining percentage was attributed to nonshared environmental effects and errors in scoring individuals. Other studies, including those of identical twins reared apart, have suggested a genetic contribution to intelligence of between 40 percent and 70 percent.[40] Clearly, these results show that natural selection has had something very tangible to work with; variance in intelligence has a strong genetic basis. More fundamentally, evolution by natural selection is *impossible* without strongly heritable traits. Unless there's a significant genetic component, there is absolutely no way of shifting population norms in a forward direction, or any other direction, as the generations move through time. A fossil record of increasing cranial capacity, in turn, makes clear that this genetic base has changed profoundly over the last 2 million years.

## HOW THE HUMAN BRAIN WAS BUILT

For many years "juvenilization" has been a popular explanation for the increase in human brain size. Since we have a flatter face that looks more like that of a baby monkey than an adult monkey, and since baby monkeys have a proportionately larger brain than do the adults, it seems logical to conclude we might simply have carried these juvenile qualities into adulthood. But such a scenario does not explain the increased growth of the brain both in the womb and in the year after birth. Biologist Michael McKinney views the larger brain of humans as the relatively simple consequence of *overdevelopment*, where growth stages, from prenatal to preadult, are prolonged.[41] The development of the mammalian brain is very conservative and differs little between major lineages, probably because any change in developmental protocols is likely to produce serious negative consequences. The overdevelopment hypothesis requires no significant changes in development beyond the addition of time. Added time means added neurons, more interconnections, and increased dendritic complexity. Because the frontal lobes (neocortex), where short-term memory and thinking are concentrated, is the last to develop, its development is the most prolonged. The frontal lobes store working memory (verbal on the left, spatial on the right) and are the areas where "executive processes" control attention, task management, and the ability to switch to other activities. Brain-scanning techniques show that the frontal lobes are where planning, creativity, and decision making have their origins. The frontal lobes appear to be the sites where a number of serious human mental ailments have their locus, including schizophrenia, depression, and attention deficit hyperactivity disorder. People with attention deficit disorder are incapable of using past experiences to control or suppress immediate distractions, with the result that they are unable to direct and coordinate their efforts to construct self-imposed, future-oriented planning. Those suffering depression can imagine their future quite vividly, but with severe negative foreboding. Among the great variety of symptoms suffered by schizophrenics, loss of contact with reality, imagined voices, and uncontrollable thought sequences are frequent. In all, these ailments give us important insights into the kinds of mental processes carried on within the healthy neocortex.

Being smarter is ultimately based on having more neurons, more dendrites, and lots more interconnections. The overdevelopment hypothesis suggests that achieving greater brain complexity was easy; it just took more time, a longer childhood, and a rich diet. This view fits in nicely with recent advances in neurobiology that have attempted to answer a vexing question: how can a limited number of genes be the basis for the development of something as complex as the human brain? Research suggests that brain growth and development is, in fact, a "Darwinian" process. Lots of neurons and even more connections develop in the young brain. Some of these are actively used by the developing child, and some are not. Those that are actively used grow and develop new interconnections. Neurons receiving little use begin to atrophy, making room for those that survive and proliferate. In this way the brain, in effect, builds itself.

## A DIFFERENT BRAIN?

Concluding his Gifford lectures in *The Human Mystery*, Nobel laureate neurologist John Eccles stated: "It is a measure of our ignorance that in the neocortex no special structure or physiological properties have been identified that distinguish sharply a human brain from the brain of an anthropoid ape. The tremendous difference in performance can hardly be attributed to a mere threefold increase in [brain cortex] modules. We have no knowledge of any quantitative development that would account for the supreme performances of the human brain."[42] Like many authors who have focused on the mind/brain *problem*, Eccles insisted on seeing a profound and unbridgeable difference between our intelligence and that of the chimpanzee. The inability to find a "special structure or physiological property" may *not* be a measure of our ignorance at all: it seems more likely that such a structure or property simply does not exist. A gradual and seamless evolutionary scenario suggests that the threefold increase in the size of our brains *was sufficient* to create those new powers that have made the human intellect unique. Terrence Deacon, in his book on the coevolution of language and brain, claims that "No innate rules, no innate general principles, no symbolic categories can be built in by evolution."[43] I disagree. All are emergent qualities created by a larger and more versatile brain.[44]

An interesting bit of evidence for the lack of an "unbridgeable gap" between our intelligence and that of our forebears is the story of Kanzi, a captive-born bonobo. Sue Savage-Rumbaugh and her associates had been attempting to train Kanzi's foster mother in the use of a keyboard and computer lexigraphs while little Kanzi amused himself in the same room. Many months of effort resulted in very little progress on the part of Kanzi's ten-year-old foster mother, and the time had come to introduce

Kanzi to the same program—"if only he could learn to sit in one place long enough," writes Savage-Rumbaugh. "Kanzi, however, had his own opinion about the keyboard and began at once to make it evident by using it on more than 120 occasions that first day. I was hesitant to believe what I was seeing. . . . Kanzi was using specific lexigrams to request and name items *and* to announce his intentions."[45] Without directed instructions, little Kanzi had learned what his foster mother could not. Kanzi had acquired these skills simply by watching what was going on around him, just as human children do. Recounting another long-term research program in chimpanzee cognition, Roger Fouts describes the acquisition and use of American sign language by Washoe and her companions, a form of communication they would occasionally use among themselves. That Washoe would "sign" to her dolls in the privacy of her own room, something she had never seen a human do, is especially remarkable.[46]

Deception is, apparently, another trait shared by chimps and humans. When Mike the chimp was asked "who ripped the jacket?" he signed *Koko*. Since Koko, another chimp, was not a possible suspect, the question was repeated; this time Mike signed a keeper's name—also extremely unlikely. Finally, with further interrogation, he signed "Mike."[47] Mike's seeming attempt to avoid condemnation by deception suggests a deep commonality in how chimp and human brains operate.

But what was the *selective pressure* that drove the continuing enlargement of the human mind? Extravagantly expensive, our brain could not have continued its unidirectional advance without the force of continuously strong selective pressures. Most authors concerned with recent human evolution have ignored the negative aspects of a massive brain and failed to appreciate the fact that without strong countervailing selection, Mother Nature could not have constructed so costly a device. Today, chimpanzees have brains only a little larger than their ancestors of four million years ago. Why didn't they continue elaborating their brains the way we have?

In 1963, Ernst Mayr commented: "The most astounding phenomenon of human evolution is the rapid increase in brain size during the Pleistocene [Ice Age]. . . ."[48] And in a 1975 review of primate brain evolution, Leonard Radinsky wrote, "One of the most fascinating unsolved problems in mammalian evolutionary history is discovering what selective pressures were responsible for . . . [the] recent and unique increase in the relative brain size of our ancestors."[49] Without this increase in brain power, our planet would not now be the home of a species pondering the cosmos. Let us next look more closely at the various selective forces that may have contributed to so rapid and so costly an escalation of human intelligence.

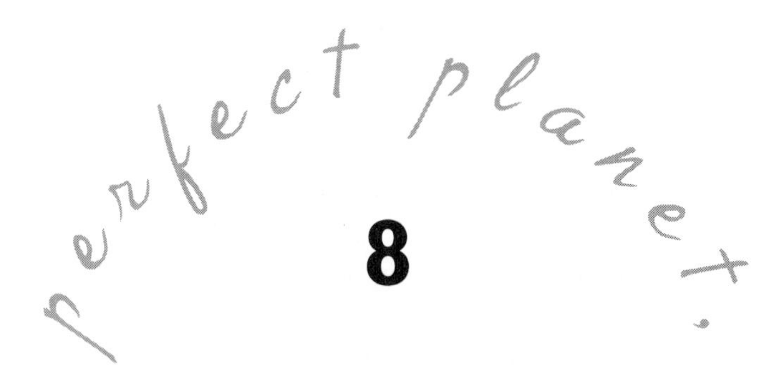


# 8

# BUILDING THE WORLD'S BEST BRAIN

clever species

Today we humans have gained a unique understanding of the universe around us. Modern science and technology have been essential elements in this advance, but without our extraordinary brains these achievements would not have been possible. The final steps in building our modern mind began well after our ancestors became two-legged. Our earliest ancestors of this period are now assigned to *Homo habilis*, the "handy man," dated at about 2.2 million years ago. This stage, the first member of our genus, was contemporaneous with stone tools that were clearly the product of human construction. Many of these tools might not have been recognized in the fossil record except for their being clustered in groups far from the strata that produced them and often associated with the bones of large game. Most early stone artifacts are nothing more than hard, fine-grained stones chipped along their narrower side. In fact, these stones may not have been the tools themselves. Thin sharp-edged flakes produced by the flaking process were probably the most often-used tools, the largest serving as knives and scrapers and the smallest used like razor blades. Such stone tools would have been effective in cutting through thick hide and tough muscle. The earliest stone tools, dated at 2.5 to 2.6 million years ago, indicate a level of dexterity in the use of the hands well beyond that of any living nonhuman primate.[1] More significantly, this early date for the appearance of stone tools implies that they were fashioned by australopithecines, *predating* the documented (but quite arbitrary) origin of our own genus, *Homo*.[2]

In modern people, 17,000 nerve fibers innervate the wrist, transferring the information of touch, texture, pliability, and thickness, while enabling us to hold, fashion, carve, delineate, and draw. Our hands and wrists are at the ends of highly mobile arms and loosely articulated shoulders. No other animal on Earth has such flexible and versatile forward appendages. They make possible not only the careful construction of tools but also the effective use of those tools in a wide variety of applications.[3]

Smashing larger bones to get at the rich marrow within was one important nutritional resource that stone tools made accessible. Stone tools could also help fashion sharply pointed sticks to dig tubers out of hard dry soil. Stone flakes could have been used to cut the tubers or tough meat into smaller, more easily chewed pieces. All these activities may help explain the reduction in the size of our molars as time progressed. In contrast, our hominid cousins, the robust australopithecines, who shared the savannas and woodlands with *Homo habilis* and early *Homo erectus*, developed huge jaws and big teeth, perhaps to chew on those same tough tubers.

Modern human hunting-gathering peoples expend considerably more energy than other primates in their daily activities. Large size, broad foraging range, and high-quality diet are all part of a pattern of increased energy consumption. Higher energy expenditure must be balanced with either more food acquisition or with higher-quality food. A larger brain requires more energy, and the quickest way to satisfy this need is to increase the amount of animal protein and fat in the diet. Some plant foods, such as nuts and pulses, may be rich in oils and proteins, but they are generally rare or available only seasonally. William R. Leonard and M. L. Robertson estimate that the gradual change in lifestyle from the australopithecines to early humans required up to 50 percent more total energy intake.[4] A portion of this increase was due to the larger size of early humans, but more intense foraging over larger areas was another

Sunset over Lake Michigan. Without the Sun's dependable warmth, an insulating atmosphere, a zesty spin, and lots of water, planet Earth could never have become the home of so many complex and diverse living things.

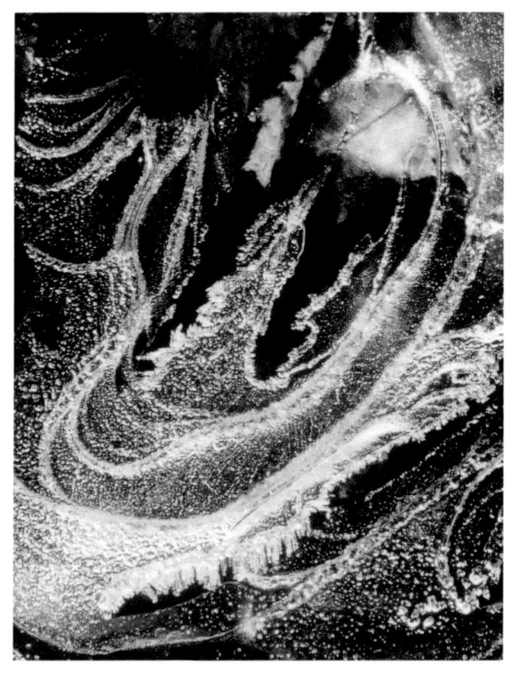

Ice formation along the edge of a brook. Only a few variables form a great variety of ice patterns along stream margins in winter. In a similar way, "hidden rules" of creative order may have helped initiate the origin of life, but at a much smaller and more complex molecular level.

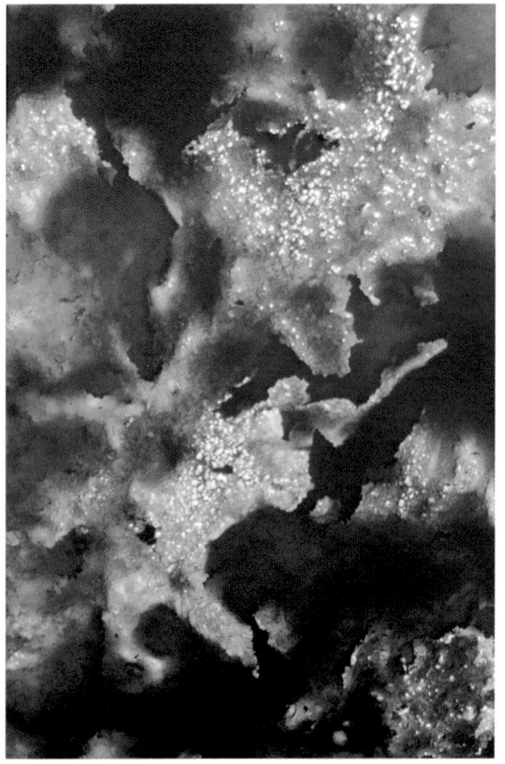

Blue-green bacteria and other thermophiles in a Rift Valley hot spring, Afdem, Ethiopia. These ancient lineages flourish in very hot water (above 150°F or 65°C), in which more complex forms of life cannot survive.

Leaf miner in an oak leaf. Eating and growing within the tissue of a single green leaf, this little caterpillar will soon pupate to become a small moth. Capturing the energy of sunlight, green chlorophyll is the primary energy source for the great majority of living things.

Wild rose with hover fly. Beginning over 100 million years ago, flowering plants and their animal pollinators formed new symbiotic relationships. These partnerships, together with a great variety of growth forms, allowed flowering plants to form landscapes richer in species than any that had come before.

Frost's bolete in oak-pine woods, western Michigan. Fungi are a separate kingdom of life. They are especially important in breaking down and recycling organic matter. In addition, many fungi form associations with roots, providing valuable soil nutrients to plants in exchange for energy-rich sugars.

Leopard frog, northern New York.
An environment filled with many hungry predators has made cryptic protective coloration an important adaptation for many animals. Disruptive coloration makes it more difficult to distinguish this frog's image from its surroundings.

Peacock displaying. Selection by choosy peahens seems to be the only reasonable explanation for such a resplendent display. In addition, any male that can carry this burden around without ending up as some carnivore's dinner has got to have good genes.

Cloud forest, Costa Rica. Cool and misty, tropical montane forests may be the world's most species-rich ecosystems. Epiphytes growing high on tree-limbs add significantly to the diversity. But despite the numbers, these moist evergreen forests support only a few species of large mammals.

Dire Dawa River, eastern Ethiopia. Acacia savanna and thornbush may number only a fraction of the species found in evergreen forests, but this drier vegetation often supports many species of larger mammals. These habitats played a major role in the early evolution of the human lineage.

Maturing head of sorghum, eastern Ethiopia. First domesticated in Africa, sorghum is well adapted to areas with short or unreliable rainy seasons. Increase in seed numbers, seed size, and seed retention at maturity were changes brought about through plant domestication by early farmers.

Highland farmer, eastern Ethiopia. Evergreen forests in the mountains have been replaced with fields of barley, wheat, sorghum, and other crops. These cooler highlands border the Rift Valley, south of the Awash River valley where Lucy once foraged three million years ago.

likely factor. Katherine Milton has recently made a compelling case for the conclusion that "the incorporation of animal matter into the diet played an absolutely essential role in human evolution."[5]

## THE *HOMO ERECTUS* STAGE

After *Homo habilis*, the next significant stage in our historical odyssey is found in an important series of fossils generally placed under the name *Homo erectus* (see Figure 2 on page 185). As one might expect, the earliest fossil fragments of *Homo erectus* are difficult to distinguish from the latest fragments assigned to *Homo habilis*. Coming onto the scene around two million years ago, *Homo erectus* was the first member of our genus to leave Africa. This was the stage first recognized by the name *Pithecanthropus erectus*, based on a cranium discovered in Java in 1891 and once publicized as the "missing link." The Latin name *Pithecanthropus erectus* means "upright ape-human," but the generic name is no longer used and these remains are now placed within our own genus, *Homo*. The famous Chinese fossils called Peking Man are also members of this stage.

A remarkable discovery along the western shore of Lake Turkana in Kenya has given us a much better picture of what early *Homo erectus* looked like.[6] This superb find, fragmented in mud 1.6 million years ago, took four field seasons to excavate from what is now solid rock. Most of us have little appreciation for the sweat and toil required to discover and unearth these relics of our ancient past. In this case, the work was well worth the effort, recovering more than 80 percent of the skeleton of a male between eleven and fifteen years of age. A remarkable aspect of the "Turkana boy" is that, had the young fellow achieved adulthood, he would have stood *at least* six feet tall. Also, he had the long-limbed body proportions characteristic of many tropical people of today. His legs, proportionately longer than those of earlier fossils, gave him a longer stride, providing a more efficient gait for walking longer distances. From his skull, it is estimated that his adult brain volume would have been 900 cc, a considerable increase over *Homo habilis*'s average of 600 cc.

Around 1.5 million years ago, *Homo erectus* developed a more elaborate stone tool technology. These more varied tools, referred to as the Acheulean industry, included thick disk-shaped "hand axes," some of which may have been used as thrown missiles. By 800,000 years ago, there is good evidence for *Homo erectus* bringing rocks from sources several miles distant from the campsite to fashion stone tools. Close to a half day's trek, such activity required advance planning. Also, by this time, some populations were consistently using fire. Along the same lines, a discovery reported in March 1998 documented stone tools on the island of Flores, southeast of Indonesia, in strata reliably dated at about 800,000 years ago.[7] What's extraordinary about this find is that, to get to Flores, these ancient people had to build some kind of navigable raft to propel themselves across a strait of open ocean that was between ten and twenty miles wide. Such abilities imply far more cultural sophistication for the *erectus*-stage people than the meager record of simple stone tools had previously indicated.

As time progressed, the fossils exhibit a transition from *Homo erectus* to "archaic *Homo sapiens*" descendants about 500,000 years ago. These finds, including the European fossil called *Homo heidelbergensis*, have brain volumes estimated to be in the 1,100 to 1,300 cc range. Despite a gradually increasing cranial capacity, *Homo erectus* was remarkably stable in general morphology, and, after initiating a new tool kit, used that technology for more than a million years.

Trying to piece together the path of human evolution over the last 4 million years is difficult because of a sparse and fragmentary fossil record. Lucy and the Turkana boy are rare exceptions. A bit of a braincase, part of a jaw from a different locality, and scattered isolated teeth are the kinds of materials paleontologists usually have to work with. This problem has been exacerbated by a tradition of anointing virtually every significant fossil with a new scientific name and populating the last 3 million years with almost a dozen "species" in a variety of genera.[8]

For example, if we allow a fair amount of morphological variation within our circumscription of *Homo erectus*, then something like the Bodo skullcap from Ethiopia, dated at 600,000 years ago, becomes part of *erectus*. But some scientists insist that this object belongs with *Homo heidelbergensis*, inserting an additional speciation event into our ancestry. Such taxonomic decisions may impose a punctuational interpretation on recent human evolution and obscure more general gradualistic trends.[9] For a general discussion, it seems best to think of *Homo habilis*, *Homo erectus*, and *Homo sapiens* as a 2.5-million-year series of chronospecies.

The fact that there is great variability among our ancestors over time and space should not surprise us; look at the variety of colors, shapes, and sizes found in *Homo sapiens* today. Why shouldn't the history of our lineage have been similarly complex over the last 2 million years?

## THE NEANDERTHAL PEOPLE

A central focus over many decades of human evolutionary studies has been the fossils assigned to *Homo neanderthalensis*. Living from about 250,000 to 30,000 years ago, the Neanderthals have always been something of an enigma. Their morphological features are quite unusual, with strong heavy bones, a strongly forward-angled face, and a cranium that is broad and rounded when viewed from behind. The Neanderthals retained a number of features of their *erectus* ancestry, including prominent brow ridges, an angled forehead, thicker bone structure, a poorly developed chin, and large upper and lower jaws. Wide-ranging research has shown the characteristic Neanderthal morphology occurring only in Europe and nearby areas of westernmost Asia.[10] Perhaps best considered a subspecies, *Homo sapiens neanderthalensis* appears to have originated from *erectus*-like precursors and remained somewhat isolated in Europe and adjacent Asia. Both male and female Neanderthals shared a heavily muscled skeleton, and together with their unusual limb-to-trunk ratios, suggest a body adapted to a very cold climate.[11] After surviving the harsh climates of Ice Age Europe for more than 200,000 years, Neanderthals gave way to a new group of people characterized by a lighter, more slender (gracile) skeletal structure and a flatter face with a higher, more vaulted, cranium. These were people like ourselves, often referred to as anatomically modern humans (or *Homo sapiens*).

The geographic isolation and distinctiveness of Neanderthal morphology has been highlighted by the recent analysis of a segment of mitochondrial DNA retrieved from a 30,000-year-old bone. While modern humans differ from each other on average at only 8 positions on this 379 base-pair sequence, the Neanderthal DNA differs from modern humans at 25 positions. Two conclusions are inferred from this difference. First, it suggests a separation of Neanderthals from their *erectus* forebears by about 600,000 years ago. Second, these people may not have contributed genes in any significant way to modern human populations. In summary, the Neanderthals are now thought to be an example of the evolution of a very distinctive regional subspecies in Ice Age Europe that became extinct. While becoming extinct is the norm rather than the exception in the history of life, the unpleasant truth is that this extinction event occurred shortly after the arrival of a rather different and more culturally versatile people.

## MODERN "GRACILE" HUMANS

The evidence distinguishing the last stage in human evolution is both skeletal and cultural. The skeleton of modern *Homo sapiens* is more slender or gracile; that is to say, not as heavy-boned as the skeleton of earlier people. The cranium of modern humans rises more steeply above the brow ridge and is more arched at the top; the brow ridge is no longer so prominent. The jaw bone is less massive and usually has a protruding chin; the teeth are smaller (see Figure 2 on page 185). These changes seem to have originated first in Africa and then, later, became evident in other parts of the world.

The heavier bone structure and the more massive musculature of Neanderthals were likely important to surviving in their cold environment, something they and their ancestors had been doing for at least a million years. In contrast, when modern *sapiens* people arrived in Europe, in the midst of a glacial episode, they were characterized by a more slender body and a wider variety of cultural artifacts. These newer people are associated with the first evidence in Europe of a technology of more diverse and finely crafted stone tools, including grinding stones used for food processing. In addition, they consistently used antlers and bones in ways that earlier Europeans had rarely exhibited. They made needles with eyes, probably for sewing leather, and more refined knives and scrapers. A greater variety of gouging and drilling tools was introduced; serrated harpoon points and fish hooks were crafted for the first time in Europe. Archaeological sites document that the modern *sapiens* peoples were bringing raw materials into their camps over a wider geographic range more often than Neanderthals had. These dramatic changes in both skeletal and cultural remains are documented in the transition from Middle Paleolithic to Upper Paleolithic cultures between 40,000 and 30,000 years ago in western Europe—a time when Neanderthals were replaced by the more modern people.

Interestingly, the archaeological record is much more complex in Israel, where modern *sapiens*-type skeletal material dates back as far as 100,000 years. This is the same region where Neanderthals also lived during the period from 100,000 to 40,000 years ago.[12] However, it is not clear whether these two morphologically different lineages of people coexisted in Israel at exactly the same time; there is no evidence for interbreeding or other interaction. With strongly changing Ice Age climates, demise of local populations, and movements of people in response to changes in vegetation and herbivore ranges, it is possible that the two different kinds of people did not actually interact over those tens of thousands of years. However, one observation is well supported: modern *sapiens*-type people and Neanderthals used similar cultural artifacts at this early time.

The shared artifacts are important: these earlier *sapiens* people did not have the fancy material culture that arrived in western Europe 35,000 years ago. Significantly, though the gracile *sapiens* people were present 100,000 years ago in the eastern Mediterranean, such modern *sapiens* remains are not found until about 43,000 years ago in eastern Europe, and 35,000 years ago in southern France. Their ability to migrate from the eastern Mediterranean to southern France was delayed for about 60,000 years. This long time period implies that a more tropical people were unable to move northward into a colder region, a region in which the heavier-set *erectus* and archaic *sapiens* ancestors of Neanderthals had lived for almost a million years. Perhaps the needles used to sew leather are critical elements of the story. Not until the slender designed-in-the-tropics people could fashion tightly fitting "tailored" leather garments to withstand the rigors of a cold climate were they able to move northward into glacial age Europe.

There has been a long history of viewing Neanderthal people as more primitive and much less intelligent than the more modern (gracile) people who replaced them in western Europe. The Neanderthals' heavier, more ancient skeletal features and the lack of elaborate cultural artifacts have contributed to these attitudes. More fundamental may be the fact that they had skulls that look quite different from those we're carrying around, and this alone suffices to mark them as different, inferior, and even suspect. Brain volume had reached very high levels in Neanderthals (up to 1,500 cc) well before the gracile people showed up. Rare flower-decorated burials, evidence for the support of old and disabled individuals, and limited symbolic markings indicate that Neanderthals had a complex social life.[13] The possession of a wider variety of more sophisticated cultural artifacts by the modern gracile peoples invading Europe 40,000 years ago may not have been due to better brains; after all, these were the same kinds of people having demonstrated no such cultural artifacts earlier in Israel. Instead, the displacement of Neanderthals by gracile *sapiens* peoples in Europe may have been caused by cultural advances having little to do with brain power. After all, both these peoples had acquired fully modern brains by at least 50,000 years earlier. Once humans had their onboard supercomputers, cultural differences would soon become the determining factor of who faded and who persisted.[14]

Cultural change is also documented in archaeological sites along the southeast coast of South Africa. The earlier people, archaic *Homo sapiens* (ca. 200,000 years ago), hunted those antelope that could easily be driven off cliffs or into swamps, yet they failed to take advantage of seasonal opportunities along the ocean shore. It was not until the advent of modern *sapiens* people that the archaeological sites contain an abundance of young seal bones. Living on rocky off-shore islands, seals give birth and raise their young in the same seasonal cycle each year. Abandoned by their mothers when old enough to fend for themselves, many young seals do not survive and are washed up onto the beaches at about the same time each year. These more modern people had apparently developed an annual calendar and a system of carrying knowledge forward through the years, making it possible for them to utilize this rich seasonal bonanza.[15] Skeletal remains of such modern *sapiens* people go back more than 100,000 years in Africa. They appear to represent the same stage as the modern *sapiens* people who replaced the Neanderthals and created the superb art on the walls of caves in southern France and Spain. We differ from these ancient people in no significant way; their skulls appear to have been identical to our own.

More significant from a general point of view is that human brain expansion leveled off somewhere between 100,000 and 40,000 years ago. Perhaps the high rates of maternal mortality in childbirth finally put the brakes on further brain expansion. Nevertheless, our ancestors were continuing their progressive trajectory. Humans were now practicing a form of progressive cultural evolution in which newly invented and useful acquired characteristics could be transferred directly to one's offspring and even across lineages. Evolution of the human genome had resulted in building a superb computer; cultural evolution would now elaborate the software, and, more important, make the difference between winners and losers.

## OUT OF AFRICA, TWICE

Contemporary people in sub-Saharan Africa exhibit more diversity in their DNA than all the rest of humanity put together. This diversity is slightly contracted in northeastern Africa and is further diminished as one takes samples from people from farther away. Such DNA data are consistent with the idea that the human lineage has spent more time in Africa than anywhere else. It is also in line with fossil evidence of the expansion of *Homo erectus* out of Africa about 1.8 million years ago, and a later origin of *Homo sapiens* in Africa around 125,000 years ago, followed by a second outward migration into other parts of the world.

The original claim of a genetic bottleneck, with all of modern humankind stemming from a single "African Eve" living about 200,000 years ago, was based solely on the analysis of modern samples of mitochondrial DNA. (Since mitochondria are inherited almost entirely through the cytoplasm of the mother's egg cell, it is appropriate to use a matrilineal phrasing.) Unfortunately, these "backward projections" were developed by use of genetic-mathematical models with a number of unrealistic simplifying assumptions, and the "Eve" hypothesis now seems seriously flawed.[16] Also, these studies based on modern DNA have the difficulty of distinguishing between a postulated expansion of modern humans about 100,000 years ago *and* the original exodus by *Homo erectus* well over a million years earlier. More important, any scenario claiming that all other genes were swept away by a tidal wave of genes out of Africa 100,000 to 200,000 years ago does not square with some of the fossil data. Facial bones of million-year-old Chinese fossils have some of the same unique characteristics seen in Chinese people today, characteristics not found in people or fossils from other parts of the world. Likewise, some people in New Guinea and Australia have a few cranial characteristics closely resembling fossil crania hundreds of thousands of years old. Such fossil and contemporary evidence suggests that there has been significant local continuity over the last million years in many areas. Those who insist on the "African Eve" scenario must postulate that the eastern Asian facial and cranial features arose twice: once in the distant past and a second time after the original features were swept away by modern *sapiens* genes. The chance of a double origin for such regionally unique features seems utterly unlikely.

While the "African Eve" hypothesis was based on analysis of mitochondrial DNA, newly reported findings based on genes of the Y chromosome give us a similar but more finely nuanced picture of recent human evolution.[17] Researchers working with these genes (the Y chromosome doesn't carry a lot of genes) have found an ancient marker base, similar to the pattern found in other primates, occurring in 15 percent of Khoisan men of South Africa and 5 to 10 percent of Sudanese and Ethiopian males. This marker has not been found on Y chromosomes outside of Africa, and such a pattern is consistent with our deep African origins. Interestingly, the Y chromosome studies indicate greater local differentiation than do the mitochondrial genes, implying that females have dispersed more effectively than males, perhaps as a consequence of bride exchange and raiding. More significantly, the Y chromosome studies indicate that at least one characteristic arose in Asian populations and was carried back to Africa. This is strong evidence for widespread gene mixing. These data picture human populations as having been dynamic, with multidirectional gene flow over tens of thousands of generations. Although there seems to be no question that some *sapiens* genes and skeletal traits, first documented in Africa, did move out over other areas of the world or were developed in parallel, it is utterly unlikely for resident populations all over the world to have vanished in advance of an onrushing wave, or for all to have been butchered without survivors. If we have learned anything in the history of life, it is that the real story is usually far more complex than anyone had at first anticipated.

But why should a *second* major evolutionary advance have occurred in Africa? The first was simple enough: our lineage originated in Africa. The second major innovation, the origin of anatomically modern "gracile" people, probably occurred for another very simple reason: for a very long time, Africa supported far larger numbers of people than any other region of the world. Also, modern analyses of climate variability reveal many sudden climatic changes having taken place with fearsome regularity over the last million years. Faced with such sudden changes, it's probable that human populations on the continent of Africa suffered fewer losses than those living in the more stressful subtropical areas of Eurasia. Despite a meager record, the bones suggest that the transformation to "modern" humans first occurred somewhere in Africa. But why there?

Why should these African people have become more slender and "gracile?" After all, *erectus*-stage people and the "archaic *sapiens*" people had been quite heavy-boned for a very long time. The explanation may be simple: natural selection tends to get rid of anything not particularly useful. Because the vast majority of mutations delete or degrade function and structure, maintaining a structure requires strong selection for those functions and structures. If selection pressures had been relaxed, bones and their associated muscles would have become more slender in modern humans, argues anthropologist C. Loring Brace. He points out that "for a period of nearly 100,000 years stone points used as projectiles occurred *only* in sub-Saharan Africa" [emphasis added].[18] With a more effective hunting technique, he argues, these particular humans no longer needed heavy bone structure and massive muscles to bring down their quarry. The new spear technology could bring down the quarry at a safer distance; it did not require the same strength as close-engagement hunting. With reduced selection pressures, skeletal robustness declined and tooth size continued to diminish (the latter perhaps also as a result of cooking foods). As these new cultural advances were adopted elsewhere in the world, other people would also become more "gracile," argues Brace. Eventually, almost everybody except the Neanderthals became slender.

## ONE SPECIES OR SEVERAL?

Widely ranging paleontological field work has made clear that the more heavily built archaic *Homo sapiens* and later, more gracile, *Homo sapiens* peoples flourished in parts of Africa and Europe at precisely the same time that people with more *erectus*-like features were still living in Southeast Asia. This is consistent with earlier but controversial findings claiming that a few fossils from Australia retained some *erectus*-like cranial features as late as 40,000 years ago.[19] Having people with *erectus*-like traits in Southeast Asia and Australia at the same time that those traits had been lost in other parts of the world should not be surprising. The problem these "late *erectus*" fossils create for human history is that they paint a much more complex picture, which confronts us with a question we discussed earlier: how do we define species?

Paleoanthropologist Ian Tattersall argues that there may have been four or five different species of the genus *Homo* in the last million years.[20] In his book *Children of the Ice Age*, Steven Stanley punctuates our history with separate "species" coming into being and going extinct with remarkable synchrony.[21] Stanley pushes *Homo habilis* back into *Australopithecus*, anoints *Homo rudolfensis* as the first member of the genus *Homo*, views *H. erectus* as a million-year-long period of stasis, and sees *H. neanderthalensis* as a species ranging from North Africa and Europe into Southeast Asia, finally supplanted throughout its range by *H. sapiens*. He sees the genus *Homo* arise, suddenly, from australopithecines that had exhibited little evolutionary change for more than two million years. These ladderlike multispecies scenarios imply multiple speciation events with the convenient extinction of previous "static species." To someone outside the field, it seems quite clear that Gould and Eldredge's "punctuated equilibrium" view of evolution has become the current scaffold on which to array these few and fragmentary fossils. Such a view is reinforced by cladistic classification, which assumes that new characteristics appear at speciation events with the concomitant disappearance (by definition) of the ancestral species.[22] Susan Cachel points out that neither cladistic methodology nor punctuated equilibrium scenarios recognize significant variation or regional differentiation *within* species. In addition, both views seem to discount the importance of selective adaptation.[23]

Today, we recognize *Homo sapiens* to be an extraordinarily wide-ranging species with many regionally differentiated populations. While the great majority (about 85 percent) of human genetic variation can be found *within* local populations, about 10 percent of human genetic variation reflects the differences between the major geographical elements in the human family. Ten percent is not a trivial amount, as is clearly evident in how easy it is to guess the geographic origin of indigenous peoples from their appearance alone. Today, we see this rich diversity as characteristic of a single polymorphic species. But when looking back in time, many paleoanthropologists adopt a very different and narrowly typological view of "species," insisting on an ancestry populated with many separate species. And let's not forget the biological species criterion: species are supposed to be incapable of exchanging genes. Considering how completely interfertile humans from all corners of the globe are today, it seems likely that early gracile *sapiens* people and contemporary heavier built, culturally less versatile *erectus*, archaic *sapiens*, or Neanderthal people were all interfertile as well.[24]

Did the gracile *sapiens* people drive the heavier built *erectus* and Neanderthals into extinction in some areas? Probably. Fossil evidence in western Europe supports this scenario. Did *sapiens* and *erectus* exchange genes in some areas? Undoubtedly; regional morphological continuity in Asia is powerful evidence for such gene mixing. Geneticist Alan Templeton has concluded that comparative analyses do not support a treelike pattern for human genealogy. Instead, he argues for a trellislike pattern, where regional groups have moved upward through time while being continuously interconnected by gene flow.[25] Yes, some skeletal traits seem to have originated first in Africa and moved outward, but other skeletal traits indicate local continuity. More important, both classes of traits are sometimes found in the skeletal material of the same individual. This serves as a reminder that traits can travel by gene exchange to nearby groups though regional populations themselves may be relatively sedentary.

Rather than a phylogenetic tree with separate branches, perhaps we should think of hominid history as being more like the currents of a broad and shallow river, unified by continuing crosscurrents of gene flow, but with some areas having quiet peripheral eddies while a few shallow rivulets terminate in extinction. Milford Wolpoff and Rachel Caspari make a comprehensive argument for this view. They see human evolution as a continuous trellislike matrix of interconnected populations in which regional traits may have persisted while more general trends, originating in different areas or in parallel, have spread, forming the complex species we are today.[26] Their view is very different from the typological and punctuated view that chraracterizes so much of the recent literature.[27] However, and regardless of which conceptual framework proves to be the most convincing, one thing is certain: within 2 million years and throughout our range, the human brain more than doubled in size. So, back to the fundamental question: Why build such a costly and complex structure?

## ICE AGE CLIMATE AS A CAUSE OF BRAIN EXPANSION

A number of authors have suggested that changes in climate during the last 2 million years, characterized by many periodic glaciations, played a primary role in enlargement of the human brain over time. Steven Stanley claims that the initiation of the ice ages 2 million years ago created an environmental crisis for our australopithecine ancestors by reducing the prevalence of trees in eastern Africa. His scenario envisions a suddenly drier climate with fewer trees, making escape from predators more difficult, resulting in the extinction of the smaller jawed australopithecines—but not before one of their populations gave rise to a brainy new genus called *Homo*.[28] However, Stanley doesn't make clear why there should be a sudden increase in intelligence in a particular group of australopithecines, other than by almost providential good luck. Also, this very same ecological crisis didn't seem to give the robust lineage of large-jawed, big-toothed *Australopithecus* species any difficulty; they kept right on chewing tubers, or whatever, until about 1.4 million years ago, when they disappeared from the fossil record.

In his book *The Ascent of Mind*, William Calvin elaborates a detailed and intricate theory for brain expansion involving both selection for juvenilization and for slowing of body development. He believes that northern ice advances caused significant population declines, followed by rapid population expansion during interglacial periods,[29] but he does not delineate exactly what the selective force for a larger ice age brain might have been. Richard Potts, in *Humanity's Descent*, makes an eloquent argument for increased climatic fluctuations having driven the evolution of our lineage and our brains.[30] But again, he provides no precise description of specific selective effects, nor is there any explanation why humans kept getting smarter and other animals did not. Authors who see ice age climate as the major driving force in recent human evolution, I believe, simply reflect our particular culture's north-temperate and Eurocentric experiences. A tropical perspective presents a very different picture.

The present period, a warmer interglacial, is the stressful period for tropical vegetation, argues ecologist Paul Colinvaux. He points out that the glacial periods average 80,000 to 100,000 years in length, while the warmer interglacials run between 10,000 and 20,000 years. Most species must adapt to the longer and cooler periods, he reasons, and their problem is to survive the short interglacials. He also disagrees with those who postulate that the Amazon basin was much drier during the ice ages, causing areas of forest to become isolated. Recent analyses of fossil pollen show that the Amazon basin was cooler but not drier during the ice times.[31] Likewise, the notion that central African rain forests also dwindled to 20 percent of their present range during the last glacial maximum may be equally unlikely. Climate changes in eastern Africa, however, may have been more complex and more severe than in the Amazon or Congo lowlands. Temperatures during periods of glacial maxima throughout the tropics probably averaged 5 to 10°F (3 to 5°C) cooler than today. But lower temperatures also mean less evapotranspiration stress for plants, with evergreen montane forests able to descend into lower elevations, expanding their aerial extent. My experience with the distribution of highland plants in Central America and in Ethiopia suggests that moist montane forests are severely contracted at higher elevations in today's hot and dry interglacial period, just as Colinvaux maintains. Pollen evidence in eastern and northern Africa documents a moister climate 28,000 to 22,000 years ago, drier 22,000 to 12,000 years ago, and warmer and moister 12,000 to 7,000 years ago, a time when the Sahara supported a rich fauna. Finally, during the last 6,000 years the climate has been hotter and drier.[32] Based on these and other data, it seems likely that eastern Africa's climate has been highly variable throughout the last several million years.

Having done botanical field work and taught plant science in eastern Ethiopia for more than four years, I find climatic explanations for critical changes in human evolution utterly unconvincing. The campus of Alemaya Agricultural College is located at an elevation of 6,500 feet (2,000 m) above sea level and boasts an ideal climate all year long. At this elevation, day temperatures rarely exceed 85°F (30°C) and nights regularly drop into the fifties (ca. 13°C).[33] Moderate elevations on tropical mountains support the world's only true temperate zone, and many of eastern Africa's elevated highlands boast such environments. Nairobi is about 5,000 feet (1,540 m) above sea level, and Addis Ababa 7,000 feet. More important, such mountains allow plant species to migrate by seed dispersal up and down altitudinal gradients in response to climatic change. Such expansion and contraction of vegetation zones was in no way equivalent to the devastation caused by colder temperatures in the north. Ecological stress caused by climate fluctuations should have been minimal in the highlands of eastern Africa.

Another important point is that most larger mammals are wide-ranging and capable of living in a variety of different ecological settings. In eastern Ethiopia, we saw the bushbuck, leopard, and the hamadryas baboon ranging from hot subdesert thornbush along the edge of the Rift Valley, at about 3,000 feet (900 m) above sea level, to cool evergreen montane forests at 9,000 feet (2,740 m) above sea level.[34] The differences between the dry thornbush and montane forest, bridged by intervening acacia woodland and deciduous forest formations, is not just one of very different vegetation—it also gets really cold at night above 8,000 feet (2,440 m), with occasional frosts. Yet many large mammals traverse this entire range of habitats in the normal course of their lives. African buffalo, lions, leopards and elephant once roamed across East Africa's highest mountains, well above the hot lowland savannas where they are more frequently seen today. Throughout the tropics, changing ice age climates probably resulted in little more than the expansion, contraction, and the continuous transformation of vegetation zones. Earlier, an increasingly dry climate in eastern Africa may have played an important role in expanding savanna grasslands and setting the scene for the emergence of bipedal hominids, but that drying trend began more than five million years ago. Here we are focusing on the glacial oscillations of the last two million years.

In contrast to Africa, northern Europe suffered severe species loss as colder weather obliterated major zones of vegetation. Many temperate species were unable to migrate southward, blocked by a nearly continuous series of east-west ranging mountains, from the Pyrenees and Alps into the Balkans. Eastern Africa's great Rift Valley suffered no such calamity. With its north-south orientation, wide array of altitudinal zones, and varied topographic features, this part of Africa offered vegetation zones ample room to move up and down the mountainsides as the world's weather became warmer or cooler, moister or drier. Referring to the mosaic of local environments along the eastern shores of Kenya's Lake Turkana two million years ago, Derek Roe has commented: "Whatever climate or physical changes time may have brought, we may feel confident that somewhere along this section of the Rift, the human population could always have found living conditions that would have seemed familiar and amenable."[35] Surviving the ice ages in this part of the world was the least of early mankind's problems.[36]

## OTHER SCENARIOS FOR BRAIN EXPANSION

If we reject ice age weather as a major factor, we can move to other suggestions for the continuing increase in human brain size over the last two million years. One possibility: the shift to a higher-quality and easier-to-digest meat and fat diet made it easier to build a more expensive brain.[37] But eating more meat and having a smaller intestinal tract doesn't explain why there should be *continued selection* for a bigger brain. Keeping our big brains cool under a hot tropical sun was a problem for our ancestors. Dean Falk's research on blood vessel patterns on the interior of hominid fossil crania indicates that the human lineage did develop better circulation to cool the brain, something the robust australopithecines lacked.[38] Such a system of cooling the brain may have been essential for making brain expansion possible but, again, it does not explain why there should be *ongoing* selection for *increasing* brain volume.

Toolmaking has often been cited as an important selective force in the escalation of hominid brain size. A problem with this hypothesis is archaeological evidence clearly indicating that humankind's tool kit remained quite uniform for long periods of time. Our first stone tool tradition, called the Oldowan, was widespread from 2.6 million years ago until about 1.6 million years ago. Having worked with this tool kit for a million years, our lineage developed the more complex Acheulean stone tool technology, and proceeded to use these kinds of tools for *another million years*. Our brain volumes expanded continuously over this entire time span, without any clear increase when new tool assemblage came into use.[39]

The development of cooperative hunting and food sharing within an increasingly complex social organization undoubtedly provided some selective pressures for expansion of the human brain.[40] These activities included a division of labor, improved hunting techniques, transport of food to a home base, and cooperative planning based on improved communication. Hunting involves stealth, anticipating the behavior of prey, choosing alternative strategies, and, finally, quick bold attack. William

Calvin believes the muscle-mind coordination needed to throw "hand axes" as missiles in hunting was a significant selective factor.[41] Studies of modern hunter-gatherer societies indicate that, while all meat is shared, it is the best hunters that have the most meat to share (it figures), and this generosity bears subtle dividends. The ladies of the group have a high regard for such hunters and provide them special privileges when their own mates are far afield. Here's a clear case of sexual selection: better hunters father more offspring. Such behavior patterns have likely had a very long history in our lineage.

Nancy Tanner postulated that food-sharing and food-gathering activities by females played a major role in human brain expansion, with males contributing little.[42] Robin Dunbar argues, as noted earlier, that language arose as a substitute for the time-consuming grooming so important for social interactions in other primates.[43] Terrence Deacon suggests that our ability to deal with symbolic concepts may have originated in rituals representing necessary social behavior in symbolic form, helping maintain allegiance, fidelity, and reciprocity within the group.[44] In his book *Chimpanzee Politics*, Frans de Waal makes clear how chimps use their intelligence and social skills in deciding "who gets what, when and how."[45] Similarly, John L. Locke argues that language arose as a means of promoting rank within the group.[46] Complex social interactions are correlated with large brains in all the higher primates, and these undoubtedly reached new levels of importance and subtlety in our lineage. Matt Ridley suggests that it was sexual selection—males and females looking for brighter, more entertaining mates—that caused our species to develop such large brains.[47]

All these explanations for brain expansion are unsatisfying for a simple reason: they fail to suggest costly, possibly lethal, consequences for failure to advance. The human brain is very expensive, both in lives lost during birthing, extended period of maturation, and in continuous energy demand. Countering such high costs requires a continuous and severe selective force. Neither hunting, foraging, nor social interactions is the likely medium for such severe selective effects. Wolves and lions are very effective social hunters, and they have not gotten much smarter over the last few million years. Baboons are quite social and many live a savanna lifestyle not very different from early humans, yet they haven't undergone unusual brain increases. Perhaps something else was going on with hominids. Recalling the studies of Vermeij, we would expect an intensively competitive arms race as the only likely selective factor to propel such a uniquely dramatic evolutionary episode of progressive escalation. But who might the antagonists be in an arms race continuing unabated for more than two million years?

## WHY HUMANS GOT SO SMART

Beginning in the last century, many theories of human evolution postulated that the expansion of human intelligence was the inevitable consequence of nature's progressive tendency to achieve perfection—or the unfolding of what they considered to be God's purposeful plan. These progressivist, religious, or teleological attitudes played a strong role in paleoanthropological thinking until the late 1940s. Finally, with abandonment of the underlying idea of an inherent progressive evolutionary drive, scientific arguments shifted to the adaptive steps that might be driven by natural selection. Now, the ecological settings and historical contingencies of our past set the tone for the style of explanations. But here again, prevailing cultural attitudes continue to hold sway over the way in which anthropologists address the question of how our lineage managed to develop the world's most powerful brain.

Many people with strong religious beliefs will prefer the unseen guidance of a purposeful deity. Trouble is, such supernatural guidance leaves no physical evidence, suggests no testable scenarios, and remains outside the enterprise we call science. On the other hand, scientists may sometimes fall prey to the fashions and sociological imperatives common to other human activities. Anthropology is perhaps more severely afflicted with this malaise than other scientific disciplines. The irrational response by some anthropologists to Edward O. Wilson's not unreasonable suggestion—that we could learn a lot about human sociology by using information derived from animal behavior and ecology—reflected this problem. When Konrad Lorenz, respected student of animal behavior, published his book *On Aggression*, suggesting that human aggression was both useful and the product of a long evolutionary history, many anthropologists had a similarly negative reaction. Such scholars worked within an intellectual framework in which evolutionary roots and animal similarities were not considered because they saw human behavior as originating entirely from learning within a specific cultural context. This view denies the existence of human "instincts" or hereditary predispositions; it found no evolutionary precursors for our behavior because it never bothered to look for them.

Ashley Montagu went so far as to claim, "There is, in fact, not the slightest evidence or ground for assuming the alleged 'phylogenetically adapted instinctive' behavior of other animals is in any way relevant to the discussion of the motive-forces of human behavior."[48] Consistent with a view of human nature as a blank slate, some anthropologists have depicted any alternative as biological determinism, preferring to view humans as infinitely malleable.[49] These attitudes became a ruling paradigm among a majority of anthropologists and have strongly influenced thinking in the field since the early decades of the twentieth century. They are nicely dissected by Robert Ardrey in the eighth chapter of *The Territorial Imperative*[50] and attacked with greater authority and many horrific examples by Michael Ghiglieri in *The Dark Side of Man*.[51] Similarly, nonbiological attitudes dominated psychology until very recently.[52] The fifth edition of a popular introductory anthropology text contained the following sentence: "The error of deciding that a given human characteristic is biological in origin, when in reality it is strongly affected by environmental factors, is dangerous because it renders irremediable something that can be remedied."[53] With such a "danger" clearly outlined, a politically correct research orientation becomes crystal clear.

A direct result of this very humanistic, biology-free, intellectual environment is that the single most powerful and parsimonious suggestion for the driving force behind human brain enlargement, *intergroup warfare*, is almost never mentioned. Papers that have developed the warfare hypothesis, such as Roger Pitt's 1978 article in the *Journal of Theoretical Biology*, are almost never cited.[54] In *The Cambridge Encyclopedia of Human Evolution*, there is only one small paragraph on warfare. It claims that "hunter-gatherers lack the social and political organization for sustained large-scale hostilities," implying that warfare was unimportant before the agricultural revolution.[55] This same encyclopedia has no entry for "conflict." As studies of wild chimpanzees have made clear (and which we'll discuss shortly), our concept of warfare need not be restricted to "large-scale hostilities." Like the American geologists who ignored the symmetry of the Atlantic Ocean because it wasn't meaningful within their ruling paradigm, anthropologists have paid little heed to the only realistic scenario for a three-million-year trajectory of nonstop human brain expansion.

Even authors as astute as Richard Leakey and Roger Lewin devoted an entire chapter in their book *Origins* to the claim that "we are a cooperative rather than an aggressive animal."[56] They failed to see cooperation *and* aggression as two sides of the same coin. Humans certainly require close cooperation and self-sacrifice within the group to meet the challenge of a difficult and unpredictable environment. However, aggressive confrontation allows a group to defend its territory in the face of challenges by other groups. Leakey and Lewin's claim that our closest relatives "the chimpanzees and gorillas are notably nonterritorial" has been contradicted by long-term observations in the field. In the period 1970–1978, Jane Goodall and her associates documented the extermination and displacement of one chimpanzee troop as a "conquering" group expanded its territory.[57] Similar observations, over several decades and in different geographic areas, have documented territorial disputes between small groups of chimps similar to such conflicts in other social animals, and they are often deadly. In such conflicts territories may shift in time and over space, and boundaries may not be clearly defined. Humans and chimps share an unusual feature among higher primates: strong male bonding—just what is needed to defend a small group against other groups. Analyses of intergroup conflicts among modern pastoral nomadic people in northeastern Africa have shown that, in the case of drier vegetation with sparse resources, there may be no clearly defined territories as such; instead, conflicts develop over freedom of movement and access to areas with available forage.[58] Early hominids in resource-poor environments must have behaved similarly. When the dry season is exceptionally long, and there is only one water hole available for many miles around, conflict may be inevitable.[59]

Arguing for intergroup conflict as the selective force that drove the last stages of human brain enlargement is not a new idea. Darwin, in his *Descent of Man*, remarks, "From the remotest times successful tribes have supplanted other tribes." Though he devotes only a few pages to this topic (most of the volume deals with sexual selection and mate choice among animals), Darwin had it right. He also claimed that sagacity was essential to defending the troop and rearing the greatest number of offspring, that groups with the largest number of talented or sagacious individuals would increase and supplant other groups—and that fidelity and courage were essential for group success.[60] Darwin's remarks also make clear why we are, at the same moment in time, dangerous killers and self-sacrificing, cooperative people. This is not a contradiction, not today or in the past. Richard Alexander sees humans as having become "trapped" by runaway intergroup conflict, which demanded close cooperation within the group at the same time that it promoted antagonistic competitive attitudes toward other groups.[61] Human brain expansion provided solutions to acute social challenges both within the group and between groups. To ensure group survival, coherent social norms within the group would integrate behavior for the common good and allow the group to meet the challenge of other groups.[62]

Focusing on the significance of intergroup aggression in no way diminishes the argument for "morality" or the golden rule having deep evolutionary roots. Doing unto others as you would have them do unto you is "reciprocal altruism" in the jargon of animal ethology, and it has been clearly demonstrated in all the higher primates. Animals able to survive only in tight social groups must have altruistic relationships for the group to remain cooperative and effective. Strong male bonding characterizes all human societies. It is a critically important behavior for group cohesion and defense. Human beings can survive and reproduce *only* in tightly structured social groups. The birth of helpless infants and many years of parenting require the kind of continuous provisioning, defense, and nurturing that only a larger group can provide. Humans cannot survive solely as nuclear families; loss of a parent by disease or accident requires immediate and continuing foster care. As philosopher Mary Midgley observes: "The long, helpless infancy which is needed to develop an intelligent adult absolutely requires a background of loyal, self-denying, cooperative elders."[63]

The deep roots of reciprocal altruism are discussed in *Good Natured: the Origin of Right and Wrong in Humans and Other Animals* by Frans de Waal.[64] A longtime student of chimpanzee behavior, de Waal makes a strong case for the importance of "good" behavior and reconciliation in primate social interactions, and he sees these as the natural background from which our own notions of morality arose. In contrast to de Waal's book, *Demonic Males: Apes and the Origin of Human Violence* by Richard

Wrangham and Dale Peterson focuses on violence by males in primate societies.[65] These authors, who have studied wild chimpanzees, argue that similar patterns of violent behavior in both the common chimpanzee and in humans are due to natural selection and evolutionary continuity. By focusing on the aggressive side of human behavior, their book makes a useful and interesting contrast to de Waal's *Good Natured*. There are no fundamental disagreements in these differing theses. Both altruism and violence are important behavioral norms for creatures living in small groups. Violence and aggression can be very useful, especially when it comes to taking over a neighboring group's resource-rich real estate.

For more than 4 million years, we and our ancestors had been living in groups averaging fewer than one hundred individuals, at campsites less than half an acre in area. We have been creatures living in small packs for a very long time—and we are still "pack animals." Have you ever met anyone who identified himself as a North American, a European, or a citizen of the world? The need to identify with a special tribe or specific group appears to be rooted in our psyche. Closely bound together with this strong group identification comes a deep distrust and easily aroused antagonism for all those who are not members of "our group." We've all been told how important self-esteem is for personal fulfillment; what better way to achieve self-esteem than to claim that your group is far superior to all the others? The heritage of being a pack animal has saddled us with a deeply dichotomous morality: us and them. Killing members of our own group is murder, but killing members of other groups is the fastest way for a male to gain social prestige. In his book *Dark Nature*, Lyall Watson points out that the Asmat people of coastal New Guinea refer to members of other groups as "the edible ones," while the Munduruсu head hunters of Amazonia call outsiders "fair game."[66]

A well-documented transformation of members of a group to members of a separated splinter group was observed among the chimpanzees of Gombe. Once the splinter group formed an independent troop, they became identified as foreign; they were shunned and hostilities soon developed. After three years, this ongoing conflict led to deadly force.[67] Researchers at Gombe had witnessed the genesis of warfare! This splitting of a population into two hostile elements is similar to what Irenaeus Eibl-Eibesfeldt has called "pseudospeciation." "Cultures mark themselves off as if they were different species," he wrote, "describing themselves as human, while all others are dismissed as non-human or not fully equipped with all the human values."[68] The reason this tendency occurs in so many social species and is so widespread within our own species is probably simple: it has survival value for ourselves and those who share our genes. Love is for those close to us and those closely related. Self-sacrificing altruism is restricted to the immediate group; it does not apply to "others." When climate deteriorates, resources become limited, or populations escalate, intergroup conflict is inevitable. And when that happens, we are preprogrammed to make sure our genes will survive.

## INTERGROUP WARFARE AS SELECTIVE FORCE

A species, whether baboon or early hominid, really doesn't need a huge brain to dig for tubers, share food, or fashion simple tools. In his book on evolutionary escalation, Vermeij claims that "competition should be an especially potent agency of selection for animals whose tolerance of starvation is low."[69] The costs of building and maintaining a bigger brain made humans *more* susceptible to starvation. Defending good territory or moving into better territory is central to avoiding starvation, and this is where intraspecific competition is most likely to occur. "Now dominating his environment, man had no serious adversary to face other than his own kind. Direct intraspecific strife—mortal strife within his own species—henceforth became one of the principal factors of selection in the human species," wrote Jacques Monod in *Chance and Necessity*.[70] Similarly, Edward O. Wilson concludes: "The evolution of warfare was an autocatalytic reaction that could not be halted by any people, because to reverse the process unilaterally was to fall victim. A new mode of natural selection was operating at the level of entire societies."[71]

Rare or sporadic, deadly conflict is the only likely source of continuous selection pressure for nonstop brain expansion throughout the huge geographic range humans have occupied over these last two million years. The effects of the ice ages were very different in different parts of this range; the effect of nasty neighbors was identical. The warfare scenario also helps us understand the rapidity of migrations out of Africa, first by early *Homo erectus* and later by the gracile people we call "modern *sapiens*." Most folks are not eager to leave their homeland; they need to be pushed. Over thousands of years, there were always other groups pushing. The warfare scenario is also useful because it applies to all but the most isolated or peripheral human populations. Was it a coincidence that human skulls became thinner in Africa shortly after spear points became a part of the archaeological record? Instead of beating each other over the head with clubs, our ancestors were now spending quality time spearing each other. Except for parasites and pathogens, other groups of human beings have always been the most dangerous elements of our environment, regardless of the climate or the continent.

Group-versus-group competition provides a meaningful scenario in regard to Jerison's requirement for a dynamic force continually selecting for increasingly sophisticated language ability. Any group whose members can communicate more rapidly and more precisely becomes a more effective fighting unit. Because this competition is between groups of the same species, it allows for the spread of superior mental, organizational, or technological traits throughout the species. Females are usually spared in such conflicts, producing additional offspring for the victors. In the biblical verses Numbers 31: 17–18, Moses, after vanquishing the Midianites, instructs his warriors: "Now therefore, kill every male among the little ones and kill every woman who has known a man by sleeping with him. But all the young girls who have not known a man by sleeping with him keep alive for yourselves." By getting rid of all males and all females who might be pregnant, this proclamation expands baby production with only a modest infusion of foreign genes.[72]

Though archaeological evidence from earlier periods is sparse, a few cases provide strong support for the intergroup conflict scenario. Recent excavations in a Spanish cave found human skeletal fragments mixed together with the bones of large herbivores in a deep pit. The skeletal materials appear to be between 325,000 and 205,000 years old.[73] Both herbivore and human bones show evidence of having been dismembered and scraped clean by the cave's human occupants. Only 4 percent of the bones show tooth marks of large carnivores, indicating that these late-*erectus*-stage people were primarily hunters, doing a bit of scavenging on the side. Some of the bones exhibit hammer stone fractures, evidence for the extraction of marrow. The important point of this finding is that the herbivore bones and the human bones had been used in *exactly the same way*—for food. The 600,000-year-old Bodo skull from Ethiopia has scratch marks on its surface and within the eye sockets, indicating the flesh had been carefully scraped out. Recent analysis of remains from a 100,000-year-old Neanderthal site in France also provide strong evidence for cannibalism. Although sparse, the evidence is straightforward; not only did ancient human groups compete, some of them ate each other.

Among social animals, aggression plays important roles in gaining access to food, maintaining individuality, competing for mates, and forming a stable within-group hierarchy. More important, aggression is absolutely essential in defending and expanding group territory. Despite the reluctance of anthropologists to acknowledge or discuss warfare as a driving force in hominid brain expansion, I believe this explanation remains the most compelling. Nancy Tanner discounted warfare theories as simply a reflection of our Western imperialist society. But how did Japan behave after becoming an industrial power in record time? And what were the Incas, Mayas, and Aztecs up to throughout most of their history? Western imperialism differed little from what *Homo sapiens* has been doing throughout recorded history. Arguing that the "*sacralization* of war" by so many human cultures is rooted in our primordial fear of being preyed upon by large carnivores, Barbara Ehrenreich makes little mention of interclan warfare and ignores territoriality. She believes the capacity for defensive solidarity arose in the "prehistoric war against animal predators" and points out that the passions of war include "courage, altruism, and the mystical sense of belonging to something larger than ourselves."[74] These same "fine passions" are no different from those we display when, rather than defending ourselves from wild animals, we are busy attacking our neighbors. For most human males, and for a very long time, warfare has afforded the ultimate measure of what it means to be a man.[75]

In *Descent of Man*, Darwin used the phrase "never ceasing wars of savages." Two world wars, the Holocaust, and recent events in Cambodia, Somalia, the Balkans, Burundi, and elsewhere make it clear that we are all "savages." Surf your TV channels; you'll have no difficulty finding murder, brutality, and mayhem. Check out the latest rioting; it's mostly males. Serbs, Somalis, or Storm Troopers don't need special incentives to kill their opponents—it comes naturally. Whether at war or in peace, ethnocentrism coupled with very negative stereotypes of outgroups is characteristic of *all human societies*, large and small.[76] Recent conflicts in

Somalia are especially instructive in this regard. Somalia is the only nation in sub-Saharan Africa with a single language and a single religion. Without color, language, or religion to distinguish each other, Somali conflicts are based on clan affiliation, and clans are defined by kinship. Because of the centrality of kinship, interclan warfare has a strong genetic component. Here is the driving force for directional evolution.

A number of biologists have claimed that natural selection is operative only at the level of individuals and their reproductive potential. Such scholars have been severely critical of the notion of group selection, but that consensus was based on a very narrow view of selection. Elliott Sober and David Wilson illuminate this question from the point of view of selection at multiple levels in their book *Unto Others: The Evolution and Psychology of Unselfish Behaviour*.[77] Indeed, small human groups may be one of the best examples of group selection (ant and bee colonies have been the favorites). By sacrificing yourself for close relatives, you can enhance the survival potential of at least some of your genes. Internal cooperation and self-sacrifice provides the group with a tightly defined individuality which is often tested against other groups. Anthropologists Jennifer N. Davis and Martin Daly argue convincingly, I believe, that competition between groups played a major role in strengthening family stability and cooperation within groups.[78] Because of cultural innovations, different groups may possess very different competitive or reproductive abilities. And just as individuals have inherited traits that promote individual survival, groups exhibit social norms having contributed to their survival as a group. On a global landscape, small groups of humans became an additional class of "individuals" that selection would determine as fit or unfit.

Robert Ardrey wrote that man "is unique only in his capacity for getting himself into troubles that for other species nature would be compelled to provide."[79] My argument here is that humans got really smart because we invented troubles for each other that nature *did not provide* for other species. By constant intergroup or interclan warfare, we created a unique evolutionary arms race and escalating feedback loop within our own species. This conforms to what Richard Dawkins and John Krebs called a "symmetric arms race," in which the parties are getting better and better at doing the same thing—as opposed to asymmetric predator-prey arms races.[80] They suggested that symmetric arms races were likely to occur between members of the same species, consistent with interclan warfare scenarios. Improved language abilities would have been continuously selected for in such an environment of ongoing deadly conflicts over countless generations. No other process seems likely to have created the extravagantly expensive but exquisitely versatile computer that sits between our ears. This scenario also deals effectively with the claim that "human intellect and language are an embarrassment for evolutionary theory, because they are vastly more powerful than anyone can account for in terms of selective fitness."[81] The only real embarrassment here is our reluctance to admit that it was a long history of intergroup warfare that elevated us to intellectual heights no other species on the planet had reached before.

But if clever chimps also display male bonding and intergroup conflict, why didn't they follow the same trajectory as humans? Why didn't chimps also build bigger brains? Two unusual human attributes suggest themselves. The first is our upright posture, freeing the forelimbs to do a lot of useful things. The second may have been our ability to make a wide range of distinctive sounds. Chimpanzees are very limited in their vocalization, and they still use their forelimbs for locomotion. With freely gesticulating hands and lots of different sounds, humans were able to develop their premier talent: language. Also, by hunting animals we gained a richer diet. And then—by hunting each other—we built a bigger brain.

In their book *The Lessons of History*, Will and Ariel Durant concluded that competition was the first biological lesson of history and that war is one of the constants of history.[82] By no means confined to recent history, intergroup warfare has been the ultimate form of human competition for at least three million years. Such competition has been the driving force behind the single most dramatic example of progressive evolution in the history of life: the expansion of the human brain (see Figure 2 on page 185).

Coming back to Radinsky's question regarding the driving force in human brain expansion over the last two million years, it seems clear that, while humans developed their large brains for a variety of reasons, the most significant was their being stalked by an increasingly clever predator that operated in small packs. As Richard Alexander has argued, "the only way to account for the striking departure of humans from their predecessors . . . is to assume that humans uniquely became their own hostile forces of nature."[83]

We'd all like to think that good things produce good results and bad things produce bad results. Unfortunately, the real world often dashes such hopes. Bad predators, bad diseases, bad climates have all bred better-adapted creatures. "Nice guys finish last!" barked baseball manager Leo Durocher to his players. Among our living relations, bonobos seem to be really nice guys. Confined to a very restricted range and with low numbers, they have clearly finished last. We humans are not nice guys, neither today nor in the past, which is why we've finished first. Clearly, this is a legacy that we must continuously strive to overcome, using the deep sense of affection and cooperation we usually reserve for our nearest kin.

At the very end of *The Origin of Species*, Darwin wrote "Thus, from the war of nature, from famine and death, the most exalted object which we are capable of conceiving, namely, the production of the higher animals directly follows. There is grandeur in this view of life. . . ."

Regardless of how far we had advanced neurologically and intellectually, late Ice Age humans were incapable of exploring the causation of disease or of building radio telescopes. A series of revolutionary cultural innovations would be required before humankind reached the pinnacle we occupy today. The first of those transformations was the development of agriculture.

# 9

# AN APPETITE FOR ENERGY

## *Revolutions in Living*

clever species

By 30,000 years ago the world's finest computer was sitting between our ears, but it was nowhere near ready to build radio telescopes. Three major cultural revolutions were necessary before humans were able to peer deeply into the heavens. For tens of thousands of years, possession of a massive brain and all the other hallmarks of modern *Homo sapiens* had not changed basic human lifestyles. Despite a more varied tool kit, exquisite artistic ability, and evidence of ritualistic activity, human communities had exhibited no major demographic change for well over a million years. As hunting and gathering bands, humans continued to aggregate in numbers between 20 and 150 individuals—not very different from a troop of baboons.

Nonetheless, big brains eventually began to make a big difference. Slowly, the human behaviors we assign to culture became more elaborate and diversified. Adaptive cultural innovations did not depend on the selective propagation of the genetic code. Now useful traits could be initiated, elaborated, and immediately shared with others. Less efficient practices could be abandoned; groups with more effective technologies would overrun those not so well equipped. In other words, culture provided humankind with a new, more dynamic form of evolution.[1] These developments had begun with the first stone tools. Later, more effective hunting abilities would bring the more gracile, modern human form into being. Finally, about 30,000 years ago, cultural diversity began to expand.

With a really fine computer in place, human brain expansion seems to have leveled off somewhere between 100,000 and 30,000 years ago. With improved language skills, complex knowledge could be transferred from generation to generation. New inventions could move between different lineages and across huge distances. Cultural evolution could be rapid and *purposefully directed.* But despite these advantages, we were still either local hunter-gatherers or wider-ranging nomadic hunters, moving to areas where game was plentiful. Around 15,000 years ago, with new hunting tools, such as spear-throwers and fluted stone spear points, human hunters became more effective; our species was beginning to make a major impact on the landscape. That impact was displayed most dramatically where modern hunters entered a new and previously isolated environment.

## HUMAN-MEDIATED EXTINCTIONS

Bands of people may have crossed the Bering Strait during the Ice Age, but those earlier immigrants have left little trace. Not until the retreat of the last ice advance, between 13,000 and 12,000 years ago, were modern hunters and their families able to migrate eastward into North America. Pollen analyses give evidence for an arctic tundra, devoid of trees and shrubs, covering the Bering Strait until about 13,000 years ago. Soon, with climate becoming less severe, willows and alders began to grow along rivers and stream margins. Once wood from these small trees was available for campfires, shelter, and tools, troops of nomadic hunters could pursue game eastward out of Asia and across Alaska into the heart of North America.[2] These were logistically organized hunting bands who could plan on a groupwide basis and move their camps in pursuit of their prey. Their spears no longer ended in wooden points; carefully crafted and firmly bound to the apex of the spear, hard stone points produced a more effective weapon. Beginning at first with distinctive spear points assigned to the Clovis tradition, a more elegant Folsom technology quickly followed. Unique to the archaeology of the Americas for many decades, these spear points have recently been joined by the discovery of a similar technology in Siberia, the probable homeland of the invading peoples. Sad to say, but for most of the larger mammals of the Americas, these newly arrived hunting cultures proved to be the beginning of the end.

Human hunters were especially effective when they encountered a new and naive fauna, as was the case in North America. Native mammoths, mastodons, horses, short-faced bears, saber-toothed cats, giant bison, several camel and antelope species, and giant ground sloths were among the victims. A wave of extinction began in northern North America and quickly swept southward to Patagonia. This carnage spared smaller creatures; rodents and dog-sized carnivores suffered no precipitous decline. Before 12,500 years ago, the North American archaeological record shows no evidence of finely crafted spear points. By 10,000 years ago, spear points are found widely over the continent, and a majority of the large mammals are gone.

Despite such damning evidence, a few scholars continue to deny human agency, claiming that "climate change" or major "faunal turnover" caused all these different large animals to meet their end during the same few thousand years. What none of these pundits have explained is why so many large animals, having survived severe climatic changes through many previous ice age cycles, should expire in grand synchrony at the end of this particular ice age? All these large animals went extinct at the *end* of a glacial period, when virtually no plants were going extinct *and* as richer more nutritious vegetation was expanding. A few of the large northern animals are still with us, including the grizzly bear, moose, elk, caribou, and musk ox; all are species that can survive in a variety of harsh and very cold environments.

Here is another example of our reluctance to accept the notion of a sudden catastrophic event or of negative human agency. Finally, the claim of a "pleistocene overkill," set forth more than three decades ago by Paul Martin, may have been vindicated.[3] By carefully analyzing growth rings within the tusks of mammoths and mastodons that died during the end of the extinction event, scientists gained new insights into the question of their demise. Growth rings within the tusk can reveal whether the animal was suffering from malnutrition and, in the case of females, how often they bore young. These analyses showed absolutely no reduction in growth rates, but the females were actually bearing young *more often*. This is exactly what one would expect from heavy hunting pressure in the presence of adequate food. Peter Ward sums it up in his book *The Call of Distant Mammoths*: "We know who dunnit. We did it."[4]

The extinction of large mammals in Europe and Asia was a slower process, but giant cave bears, wooly rhinoceros, mammoths, mastodons, and the grandly antlered Irish elk were also exterminated. Again, some scientists insist that climate change killed animals that had survived two million years of severe climatic fluctuations. Extinction of large mammals was less sudden and less severe on the continent where humans had the longest track record. The African fauna did suffer a significant loss of large mammals, such as the saber-toothed lions and giant pigs, but those losses appear to have begun more than 150,000 years ago. Having developed in concert with their evolving two-legged adversaries, the surviving African fauna may have been better able to adapt.

Australia, the island continent, experienced the arrival of humans around 60,000 years ago. By 30,000 years ago, most large marsupials were gone—giant kangaroos and a rhinoceros-sized Diprotodont among them. Again, some scientists claim that the environment of Australia during this time was deteriorating as the climate became drier, as indeed it was, and this was the major cause for the decline of large mammals. A new study has demolished this argument, using the eggshells of a large extinct flightless bird. Analyzing the ages of more than a thousand fossil eggshell fragments in several different regions of Australia having had different ecologies at the time, this study made clear that these giant birds were continuously abundant from more than 100,000 years ago up to 50,000 years ago. The last of these large birds all expired shortly after humans arrived.[5] A similar set of circumstances seems to have been in play more recently, in Madagascar, where drying trends became more severe between 8,000 and 5,000 years ago. Though some lemurs went extinct earlier, it was after humans arrived, about 2,000 years ago, that fourteen species became extinct. All were large and slow-moving, feeding at ground level in the daytime. Those lemurs surviving today are mostly small tree dwellers that are active at night. When humans arrived, both

Madagascar and New Zealand lost their giant flightless birds. The more distant islands of the Pacific, including Hawaii, were also the homes of many endemic and smaller flightless birds. Virtually all became extinct once humans reached those distant shores.

The notion that earlier cultures were especially sensitive to the well-being of native animals and carefully conserved their environments is a myth of recent vintage. Beginning more than 50,000 years ago, extinction episodes over the entire planet bear testimony to our species' hunger for the tasty caloric content of larger mammals and flightless birds. With more sophisticated weaponry, better coordinated hunting protocols, and the ability to reach isolated regions, the human species had begun to make its mark on the biosphere.

## A DIVISION OF LABOR: TWO VERY DIFFERENT SEXES

Not only do contemporary people around the world share the same general skeletal features characterizing gracile humans of tens of thousands of years ago, we also exhibit the same distinctive physical differences between males and females (called "sexual dimorphism"). This subject brings us back into the business of sex again; but let's be clear about the differences between sex and gender. Sex deals with chromosome shuffling and gene recombination to produce genetically diverse offspring, while gender denotes the characteristics distinguishing male from female. Fungi, ciliates, and some algae have sex without gender: their bodily form and their sex cells are indistinguishable; they have no males or females. Gender in animals and most plants began long ago with the development of two different kinds of sex cells. One was larger and plump, storing more energy to help foster a new life. Such cells are often sedentary and we call them the females. The other type of sex cell was much smaller and outfitted with a long, wiggly tail. Highly mobile, these little fellows, the males, would have to swim through a dangerous world in search of a female cell with which to form a happy union. Many plants, such as algae, mosses, and ferns, have sperm that must swim to the stationary egg cell. Simple and pragmatic, this two-gender system is found in many plants and animals; it has constrained our lives ever since.

A fundamental aspect of human gender differences originated with the earliest mammals, as females became responsible for nursing and caring for bigger-brained and more energy-demanding offspring. This original gender gap widened further as primate mothers were forced to spend longer periods of time with their smart and slowly maturing infants. The gap became even more pronounced as human females gave birth to helpless infants who matured even more slowly. With such compelling responsibilities falling to human females, it seems obvious that human males should have been usefully occupied in other ways.

As foraging bands of hunter-gatherers, the roles of males and females have probably been differentiated for many millions of years. Hunting, territorial defense, and physical combat are reasonable explanations for an overall heavier skeleton, thicker skull, greater upper-body strength, and generally more violent temperament among human males. Women have less robust bones and ligaments than men. Nowadays, women in sports suffer between four and six times as many anterior cruciate ligament injuries of the knee as do men.[6] On average, human males have a gripping-holding strength almost twice that of females. Males also have a slower heart rate while resting, greater blood flow volume, and a greater capacity for neutralizing lactic acid as it builds up during strenuous exercise, all characteristics of a more energetic lifestyle.

Except for gibbons living as solitary families in treetops, all the apes exhibit considerable sexual dimorphism, with males consistently larger than the females. Though size dimorphism is not as pronounced in humans as in orangs and gorillas, there is another striking difference in our own species: fat storage. The marvelous ability of human females to convert food into curvaceous deposits of stored energy has sustained women who bear and nurse children over short-term food shortages. The amount of fat in mature human males averages about 14 percent of body weight, not much different from the primate average, whereas the average for women is about 27 percent of body weight. Additional fat plays an important role in creating big-brained children, as well as providing an onboard insurance policy for tough times. With the "cost" of a full-term pregnancy calculated to be about 50,000 calories and lactation estimated to require as much as 1,000 calories a day, having an onboard larder had an obvious selective advantage for a woman's ability to produce and sustain an infant through episodes of food scarcity. On the other half of the scorecard, extra weight would have been a serious disadvantage for those who spent time trekking long distances in the pursuit of precious protein and energy-rich fat. Differentiation of the sexes in body and temperament, the product of thousands of generations of ongoing selection, was a key factor in making ours so successful a species.[7]

Just as human males and females differ in body form, musculature, and weight, there seem to be slight differences between both the structure of their mature brains and some aspects of cognition. Severe injuries to the left side of the head usually cause more serious loss of speech function in males than in females. Consistent with these data, modern imaging techniques show women using more diverse areas of their brains when engaged in speaking. Furthermore, women consistently perform better in verbal abilities, social sensitivity, and reading comprehension while males are somewhat better at map reading and recognizing similarity in complex three-dimensional images. Women use and remember landmarks while men tend to use distance and orientation to figure out where they are. Such differences are exactly what one might expect in a species in which the males have been the wide-ranging hunters and females the more sedentary gatherers. Such a division of labor makes economic sense. Digging tubers and picking berries may provide the group with a large majority of its caloric intake, but it is time consuming and demands remembering where these resources are located. "Women not only have a better memory for objects and their locations, but they naturally pay more attention to their surroundings," notes Robert Pool in *Eve's Rib*.[8] A group's survival may have depended on women's ability to find tubers in the leafless dry season, based on a keen remembrance of *exactly* where a particular leafy species was seen during the rainy growing season, many months earlier.[9]

Just as nature has fashioned two very different genders in our species, many cultures have defined the roles of men and women very differently. The nearly worldwide confinement of women to childbearing and child rearing in traditional societies is, very likely, much more than a simple conspiracy by power-hungry males. In the considerable comforts of our modern times, we forget how pestilence, famine, warfare, infant mortality, and the death of women in childbirth have ravaged human communities throughout history. Societies that did not keep all their young women in the business of producing babies seriously compromised their ability to sustain their numbers or defend themselves. Only in modern times, when most children survive and human populations are exploding, have societies achieved the freedom to offer women a wide choice of lifestyles. Also, just as traditional cultures have constrained the life of women in times past, natural selection appears to have kept variation among women more tightly constrained. And probably for the same fundamental reason: women's basic role in birthing and nurturing was too important.

## GENDER AND VARIANCE

In some quarters, discussions regarding innate biological differences between the sexes is considered part of an unacceptable acquiescence to the dogma of biological determinism. But the statistics seem clear enough: males are more expendable than females. Most obvious is the fact of the human male's shorter average life span. (This was not so apparent in earlier times when many women died in childbirth.) Today, a near-identical percentage of women die of heart disease as do men; the only difference is that women's heart problems begin about ten years later. These statistics are consistent with the finding of females outliving males in nearly all primate species for which there are accurate data. Interestingly, the very few exceptions are species in which the males play an essential role in caring for the young.

More troubling are the equally grim statistics showing how human males consistently outnumber females in autism, learning disabilities, dyslexia, stuttering, schizophrenia, attention deficit hyperactivity disorder, Tourette's syndrome, and various forms of dangerous and self-destructive behavior.[10] Those last characteristics, needless to say, do not enhance male life expectancy. In nearly all human societies, males are consistently more aggressive than females. Nearly all violent crimes against people whom the assailant does not know are committed by males. Take a good look at violence around the world on the TV news; it's mostly males. Such observations parallel those among all nonhuman primates where it is the males who are more aggressive, territorial, and vigilant. These basic behavioral differences seem to be closely correlated with different hormonal levels in all primate males and females; eight- to twelve-year-old boys have twice the androgen levels as do girls of the same age.[11] Likewise, behavioral differences appear in early childhood and continue throughout much of the life span in every corner of the globe.[12] Teenage turbulence in both sexes and violent behavior among young males are additional innate hormone-driven realities, not artifacts of culture. A wide range of individual variation within populations all over the world tempers these generalities, but it does not alter them.

An intriguing and perhaps instructive difference between males and females is found among those rare people especially gifted in mathematics, where the number of men consistently exceeds the number of women. As someone who is severely challenged by mathematics in any form, I had been puzzled by an anomaly in the statistics of mathematical abilities among young people. The anomaly is this: though males are more numerous among the mathematically gifted, many studies show that average male and female scores are *identical.* A Scandinavian study involving thousands of students recently corroborated these earlier findings but provided a simple explanation for the disparity. Just as there were more mathematically gifted males, there were also more mathematical dummies among the Scandinavian males. The average scores were indeed identical; it is male *variance* that is greater.[13] Apparently, in most earlier studies, no one spent much time checking out the underachievers. This finding is concordant with IQ tests in which there are generally twice as many males scoring below 55 and above 145 as females. All these observations are consistent with data regarding mental disabilities and a greater incidence of aberrant males in all human cultures. Greater variance in males explains these unfortunate statistics; but greater variance also means an increased likelihood for unusually gifted individuals.[14]

Though talented individuals may have had limited potential in small groups of wandering hunters and gatherers, in large settled societies gifted individuals would make significant contributions. By entering into new "symbiotic" relationships with domesticated plants and animals, we humans suddenly transformed our ability to sustain ourselves. With the invention of agriculture, daily concern for food procurement was no longer the central problem for the community. Now, human cultures could diversify in creative ways that had never been possible before.

## THE AGRICULTURAL REVOLUTION

Whether antelope, amoeba, or turnip, every living organism must meet its ongoing energy requirements. Run out of energy and you are history. Virtually all living things store energy to bide them over times of scarcity or through seasons of inactivity, but such resources are limited. For wandering hunter-gatherers dependent on wild plants and game in an environment of uncertain weather, starvation was an ever-present threat. Developing an assured supply of food energy that could be stored for unproductive seasons or bad years was the single most important cultural advance in the modern history of humankind. Agriculture was that revolution.

The exact way in which humans first domesticated plants and animals has been the subject of much conjecture and many theories. An increasing number of careful archaeological studies now suggest that settled farming villages developed in differing ways under different environmental conditions and over long periods of time. People in at least five parts of the world—the Middle East, China, Southeast Asia, Mesoamerica, and South America—developed agriculture early and independently. Two general factors have been invoked for explaining these distant but concurrent trends: (1) environmental changes at the end of the last glacial cycle and (2) ongoing cultural advancement, with small, seasonally settled communities under pressure to acquire more reliable food resources.

The earliest generally accepted date for plant domestication was thought to be about 9,000 years ago in the "Fertile Crescent" of the Middle East. This crescent sweeps northward from Israel and western Jordan through Syria and eastward across southeastern Turkey, then southeastward into the drainage basins of the Tigris and Euphrates Rivers in Iraq. Here, einkorn wheat, emmer wheat, barley, peas, chickpeas, and lentils were among the earliest of our domesticated plants.

Even earlier dates for domestication have recently been documented for rice in the central Yangtze valley of China. Here, phytoliths (minute crystals formed within plant cells) capable of distinguishing between the archaeological remains of wild and cultivated rice show a transition from a limited diet of wild rice about 12,000 years ago to a diet rich in cultivated rice by 10,000 years ago.[15] Recent archaeological work in Mexico indicates that squash domestication may have begun in Mexico as early as 9,000 years ago, 4,000 years before maize became an important staple.[16]

In the Middle East, animals may have been domesticated before the first local grains became regularly tended field crops. For hunter-gatherers traveling with captured young goats and sheep, animal domestication might have been an easier cultural innovation than learning to plant seed and till the soil. Because they lived naturally in hierarchial social groups, goats, sheep, pigs, and cattle captured as young animals required little modification to become part of a human-dominated herd. By around 8,000 years ago, sheep, goats, cattle, and pigs had become a regular part of the human community in the fertile crescent. It seems likely that these four species were first domesticated in different regions, but, once they became our partners, were quickly adopted by people living elsewhere. Pigs, like dogs, played a useful role as unpaid sanitary engineers in crowded villages. Cattle herding may even have preceded agriculture in the Sahara, before that huge area became the inhospitable desert it is today.

About 14,000 years ago, with a warming at the end of the Ice Age and increased rainfall in the Middle East, woodlands and a diversity of herbaceous plants began to expand their range into what had been drier steppe-like areas. An ameliorating climate with increasing nutritional resources allowed hunter-gatherers to increase their numbers, but then came a nasty change in the weather. Called the "Younger Dryas" in Europe, cold temperatures suddenly returned and were maintained for hundreds of years around 11,000 years ago. This must have reduced available food resources, which may have forced hunter-gatherers to find new ways to support their populations.[17]

Small settled human communities probably existed along river valleys with diverse vegetation well before agriculture became established. Here is where the discipline of soil preparation and careful propagation of plants probably first took place. But what might have inspired humans to make so drastic a change, from gathering and hunting in a free and nomadic lifestyle to the drudgery of soil preparation and the burden of having to tend field crops? Wheat and barley apparently grew in such profusion in particularly favorable sites in the mountainous terrain of the Middle East that it was possible for local populations simply to gather the wild plants over many thousands of years. However, as populations expanded, people would be forced into peripheral areas where wheat and barley plants were too widely scattered to make gathering them feasible. It is perhaps in these peripheral areas where deliberate planting and tending began.[18]

There has been a long-standing debate over whether agriculture originated first in one area and was then transmitted to the others, or whether it developed independently in more than one area. The earliest crops of the separate "centers" are quite distinctive, and most scholars now believe their origins took place independently. Clearly, agriculture developed independently in several regions of the Americas, but perhaps a bit later than in the Old World. Once established, the new life-style advanced inexorably. Comparative archaeological research has documented the gradual spread of agriculture from the Middle East westward across Europe, reaching Greece about 7,500 years ago, Italy and Romania by 6,500 years ago, France by 6,000 years ago, and northern Germany and Scandinavia between 4,000 and 3,000 years ago.[19]

Why did settled agriculturalists prevail, forcing hunter-gatherers into fringe areas of low productivity? Here is one of the few questions on which the archaeological record provides clear and unequivocal data: after the advent of settled agriculture, human occupation sites increased tenfold in area. Campsites of hunter-gatherers averaged half an acre (0.2 hectares) in area, while the first agricultural settlements covered an average of five acres (two hectares). Whereas hunting and gathering usually supported only several dozens of people, a more predictable food supply could support hundreds. By maintaining a settled village life, death rates probably decreased. A sick person, protected from the elements and asleep within a permanent abode, had a far better chance of recovery than someone forced to keep up with a wandering band.

Agriculture allowed many landscapes to support far larger human populations—whether settled farmers or wandering pastoralists with domesticated herds—than had been possible before. The demographic advantage of agriculture was simply overwhelming. In the best locations, especially the fertile flood plains of large rivers, cultivated plants and domestic animals could support thousands.[20]

## PLANT AND ANIMAL DOMESTICATION

Freed from a precarious wandering past, farming communities developed a uniquely symbiotic relationship with their crops and livestock. This innovative lifestyle could have originated only in a topographically diverse environment with a rich variety of plants. It is from among weedy flowering plants growing quickly in open sunny sites that our most important food crops arose. Such plants are especially efficient at converting sunlight into the stored energy found in fruits, seeds, stems, or roots. Of all these varied food sources, it is the seeds that have played the most important role. Seeds are the disseminules of higher plants. They contain both the embryo and sufficient energy for development of the seedling until its own photosynthetic machinery can power further growth. Seed plants had originated during the ancient coal ages, but it was the more modern flowering plants that formed a unique leaflike enclosure in which their seeds would develop. Different lineages of flowering plants elaborated this new encasement to become the succulent outer flesh of the apple, avocado, and olive, the hard shell of the pecan, the husk of the coconut, the pod enclosing peas and beans, or the outer layer of the grass grain.

Among the many plant species contributing to our nutrition, the grass and legume families are among the most significant. The grass family provides us with the cereal grains while the legume family includes the larger and more nutritious "pulses" (beans, peas, lentils, etc.). Grasses and legumes are also the major components of pasture for our grazing animals. If we insist on counting only calories, it is the cereal grains that are the most important. Members of the grass family transform more sunshine into people and livestock than any other plant group. Of the twenty most intensively cultivated plant species, eight are grasses. The small seeds of wheat, rice, maize, barley, sorghum, rye, and the millets (all are grasses) helped begin agriculture, and they continue to energize us. Rich in nutrition, low in moisture content, small, and hard, these tough little grains are eminently practical. Stored in a dry place, away from vermin, they can be kept indefinitely.

Seeds of the legumes or pulses are larger than grass grains and more nutritious, thanks to a higher percentage of proteins and lipids. Like the grains, they are hard, low in moisture, and easy to store or transport. The pulses include many kinds of beans and peas, as well as lentils, peanuts, and soybeans. Nodules on their roots, housing symbiotic bacteria, are an unusual feature distinguishing members of the legume family. Fixing nitrogen from the air, bacteria within the nodules provide available nitrogen to both the legume plant and, after the plant decomposes, to the soil in which it grew. That's the good news. The bad news is that you can't get something for nothing. The legume plant must produce special structures (the "nodules") and organelles (the "symbiosomes") in which to house its bacterial allies, as well as special proteins (called "nodulins") to help sustain them. Maintaining their symbiotic bacteria and producing more nutritious seeds requires more energy, with the result that legumes cannot produce as many pounds of seed per acre as do the grasses. In contrast to the major grains, which contain between 7 and 14 percent protein by weight, the pulses range from 20 percent protein in chickpeas to 45 percent in soybeans. By developing a good balance of grains for energy and "beans" for proteins, these crops come close to providing humans all they need to survive and grow. (The sparsity of these plants in wild landscapes required people to hunt animals in earlier times.)

Root crops have also been important staples in many cultures, especially in evergreen tropical lowlands. Although not all are roots in a technical sense, these crops include the sweet potato, cassava, taro, yam, turnip, potato, and a variety of others. Because root crops contain more moisture than the grains, they are less concentrated in energy value and more susceptible to spoilage. They are also lower in protein content, but many have the advantage of being stored by simply leaving them in the ground, to be dug up when needed. The potato, developed in the cool highlands of the Andes, is now the world's fourth most important food crop, after wheat, rice, and maize—in that order. Though historians often mention the great Irish famine, caused by the potato blight disease, few people realize that the human population of Ireland had *more than tripled* as a consequence of introducing potato cultivation to the emerald island.

Some regions of the world are distinguished in having a suite of important indigenous crops, and they are thought to have been centers of agricultural innovation. The Americas have four such areas. Mexico and Central America gave the world maize, tomatoes, grain amaranths, cacao, avocado, papaya, vanilla, and specific varieties of chili peppers, squashes, kidney beans, and cotton. The high Andes are the original home of root crops such as potatoes, oca, ulluca, quinoa (a grainlike seed) and lima beans. From lower elevations, South America gave us cassava (manioc), sweet potatoes, peanuts, pineapples, cashews, tobacco, and important varieties of cotton. Finally, and at a considerably later time, the lower Mississippi and Ohio valleys of the United States witnessed the domestication of marsh elder and the native goosefoot, as well as development of new varieties of corn, pumpkins, and sunflowers. Likewise, there were a number of important areas of local domestication in the Old World, apart from China and the Middle East. Africa south of the Sahara gave rise to pearl millet, African rice, sorghum, cowpeas, indigenous yams, okra, the oil palm, and a variety of melons. Coffee originated in the high mountains of Ethiopia. Here also is the exclusive home of teff (the world's smallest cereal grain), ensete (a banana relative with starchy base), noog (an oilseed), and distinctive varieties of sorghum and barley. Tropical South and Southeast Asia were the original locales for the domestication of rice, bananas, plantains, mangos, many citrus fruits, cucumbers, black pepper, eggplant, yams, and many kinds of beans. New Guinea and the western Pacific gave us coconuts, taro, breadfruit, wing beans, sago palm, and sugar cane.

A few plants and animals of the same species (or closely allied species) appear to have been domesticated more than once *independently*: for example, pigs in western Asia and in Southeast Asia, varieties of kidney beans and chili peppers in both Mexico and South America, and rice in several regions of Asia. Recent DNA evidence suggests that cattle may have been domesticated independently in three areas: northern Africa, the eastern Mediterranean, and India. Much earlier, dogs had been independently domesticated from wild wolves and their close relatives in several different parts of the Old World, giving rise to animals as different as the German shepherd and the Australian dingo.[21]

Unfortunately, the agricultural revolution also had negative consequences. Skeletal remains from early agricultural communities sometimes display the effects of nutrient deficiencies and dental decay; also, settled agricultural people became smaller than their predecessors. Such data document a *decline* in human health among some agriculturalists, as compared with earlier hunting-gathering peoples of the same region.[22] Deformed backbones, probably as a result of continually grinding grain or tilling the soil, indicate that the new agriculturalists had to work harder and longer to maintain their new lifestyle. Roy Porter notes that "while agriculture rescued people from starvation, it unleashed a fresh danger: disease."[23] Crowded into settled agricultural communities and living in close relationship with domesticated animals, diseases could now spread with much greater rapidity. And yet we prospered.

Grains, pulses, and root crops provided food resources with which to fend off the specter of starvation. When supplemented with wild game or the meat and milk of domestic animals, the diet could be complete and well balanced. Spices, fruits, and vegetables also provided vitamins and micronutrients needed to supplement the basic starchy staples. Some traditional starch sources (such as maize, taro, and cassava) have been denigrated by nutritionists as not providing sufficient protein, but such criticisms are unrealistic. (There are other sources for proteins.) I prefer the statement: "There is no such thing as a bad food. There are only bad diets!"[24]

Just as the wild ancestors of important agricultural plants had a set of characteristics making them amenable to gathering, manipulation, and planting in soils prepared by people, domestic animals also had to possess behavioral characteristics that made their successful domestication possible. Herding and training larger animals was feasible only with animals that lived in close social groups with a clearly defined dominance hierarchy. Genetically programmed for conforming to a social order, these species could easily adjust to the dictates of their new masters. Today, humankind maintains something in the neighborhood of 18,000 million chickens, 1,800 million sheep and goats, 1,300 million cattle, and 900 million pigs, providing us with about 16 percent of our worldwide caloric intake. Sheep and goats were probably the first important animals domesticated for food, followed by pigs and chickens. Cattle may have been originally domesticated for religous rituals and later for their milk. These larger beasts had an additional value: they turned wayside weeds into cow pies that could be used as manure, mixed with mud and thatch to make a stronger wall, or when properly dried, used as briquettes for the kitchen stove. Use of the plow and wheeled carts pulled by oxen, beginning between 3500 and 3000 B.C.E., made animal traction available for tilling the soil and hauling material goods—a major advance in our expanding acquisition of power. Domesticated a bit later, sure-footed donkeys, at home among rocky scrubland, provided transport over steep and narrow trails. Horses, first domesticated on the broad plains of central Asia, spread, not so much as an agricultural animal, but more dramatically as a new element of mobile warfare.

## THE RISE OF CITY-STATES

After its inception, agriculture spread widely and ever since has provided a reliable basis for human sustenance. The question of why agriculture and, somewhat later, large city-states developed so early and so successfully in the Middle East as opposed to other parts of the world is analyzed by Jared Diamond in his book *Guns, Germs, and Steel.*[25] He notes that Mediterranean vegetation covers a large area around the Fertile Crescent and this, together with the adjacent mountains, support a great diversity of plant species. It was from this environment that early Middle Eastern farmers obtained a rich variety of food plants, spices, fibers, and fruits.

As Diamond points out, the huge Eurasian land area is home to seventy-two species of larger herbivorous mammals, of which we've domesticated thirteen. In the Middle East, goats and sheep, followed later by cattle and pigs, provided meat, milk, hides, wool, and fertilizer. Peoples beyond the Eurasian landmass did not have such a large array of species amenable to domestication at their disposal. Though Africa has fifty-one species of large mammalian herbivores, only the wild ass and African cattle were transformed into useful domestic animals.[26] One would think that at least one of Africa's many species of antelope could have been domesticated. Unfortunately, most male antelopes mate only on clearly defined and fiercely defended territories—a trait rendering them impotent in a crowded corral. Zebras lack some of the qualities of personality that make horses and donkeys so manageable. Obviously, only those animals (and plants) with the right combination of inborn traits could become close partners with humans.

Diamond's more general thesis is significant: people of the Middle East did not develop agriculture to such high levels because they were smarter than others. Rather, they benefited from a rich and diverse biota and a huge landmass that facilitated effective east-west exchange of crops, animals, and cultural innovations. As one might expect, many crops were difficult to transfer into areas of substantially different climate or day length. Wheat traveled easily from Turkey eastward to northern China where it became a major food source, but wheat would not flourish when transported similar distances north and south. The potato was not cultivated outside the Andes until after Columbus, when it was transplanted to other cool climates. It was adopted only slowly in northern Europe, perhaps because it took time to find varieties for the longer day lengths of the northern summer. Today our markets are provisioned with crops developed by indigenous cultures throughout the world over thousands of years. Just how successful those early farmers were is reflected in a sobering fact: we have not added a single major staple to our food inventory over the last *two thousand* years.

Landscapes under the dominion of the agriculturalist grandly expanded our share of the Sun's radiant energy. Both settled farmers who tilled the soil and wandering pastoralists greatly multiplied the food resources available to humankind. But despite their success, and even with tens of thousands of small agricultural communities, large city-states and their civilized lifestyles developed in only a few localities, and several thousand years *after* agriculture had been initiated. Apparently, people gave up their autonomy only under special circumstances, joining together to become part of the larger regional social organizations we now call city-states. It is probably no accident that all the earliest major civilizations arose in strongly seasonal environments, where long dry or cool seasons reduced soil leaching, helped preserve stored food, and made farm labor available for other activities during part of the year.

Why should separate local communities give up local autonomy and come together to form a larger regional state? This has been a central question in anthropology for many decades. Early explanations calling on population pressures or climate change have proven to be unconvincing. Archaeological evidence records the beginning of states where no changes in population size are evident. And since the climate and environment have never really stopped changing, such explanations are suspect. Currently, the notion that "people are selfish and the accumulation of wealth is a primary driving force in human affairs" underlies attempts to explain the origin of large states.[27]

There is good evidence that virtually all village agriculturalists fail to reach the levels of productivity or consumption of which they are capable. Organized in larger groupings, these same villages can increase their productivity and have access to a far wider range of resources. In a "civilized" state with a system of maintaining order, a well-organized labor force, long-term building programs, and safe efficient transportation arteries, the elites and everybody else can have access to affordable luxuries not possible among small independent villages. Irrigated agriculture, requiring long-term planning, careful construction, and continual repair, is just one example of how the organized state could achieve new levels of productivity and efficiency. Longer-distance commercial interchange over well-maintained and well-protected transportation routes was especially important. Interregional warfare may also have been a strong incentive for the development of the larger organized state.

Despite the success of agricultural villages in many areas of the planet over thousands of years, less than a dozen larger states emerged to become major early civilizations. These provided a setting in which monumental architecture, religious priesthoods, social hierarchies, the art of writing, and metalworking would come into being. Often located along the shores of seasonally flooding rivers, the large "hydraulic" states brought the agricultural revolution to its full development. Unfortunately, these large societies also promoted the formation of ruling elites, condemned many to the role of subservient peasants, and inspired large-scale warfare. But a world without an advanced agricultural state was a world without priests, architects, accountants, or, more central to our thesis, metalworkers.

## LEARNING TO WORK WITH METALS

The metals we have come to rely upon have unusual qualities possessed by few other substances. Their tensile strength allows them to resist bending or breaking, or fracturing under compression. Metals are the substance of choice when one requires resistance to abrasion (as in bearings or plows), buckling (nails), twisting (drills), stretching (cables), and shattering when struck (anvils). Especially significant is the fact that some metals maintain their important qualities at a small size, as in knife edges and precision tools. Replacing tools of stone, wood, and bone, metalworking allowed the further creation of tools that would have been unimaginable in earlier times.[28]

Just as our agricultural revolution could not have taken place without a variety of nutritious foods, the beginnings of metallurgy are difficult to envision on a planet without a dynamic surface. The average proportion of copper in the Earth's crust is a measly 0.0058 percent (by weight), yet there are rock strata on the eastern Mediterranean island of Cyprus (the word means "copper") which are 2 to 5 percent copper; and rare veins of pure copper have been found elsewhere. Stony meteorites have elemental proportions not unlike those found on Earth, but none are especially rich in copper; nor has anyone ever found a copper meteorite. If that's the case, how did the rich copper ore deposits in Cyprus and elsewhere originate?

Here's another example of how the theory of plate tectonics has expanded our understanding of planet Earth. Recurring earthquakes, picturesque volcanoes, and complex mountain chains are only the more noticeable effects of the dynamics at plate margins. Colliding plates and associated subduction zones also provide the conditions for intense heating of minerals at depth and under high pressure, creating cauldrons for the recycling and concentration of uncommon elements. This is where ores, enriched by these dynamic processes, acquired high concentrations of rare and useful metals. As luck would have it (there's that phrase again), the eastern Mediterranean basin and nearby mountains of southwestern Asia were not only rich in a wide array of flowering plants, helping provide the basis for the development of local agriculture, this mountainous region is also the point at which the African, Eurasian, and Arabian plates collide. The eastern Mediterranean Sea reaches depths of over 6,000 feet not far from the copper-rich rocks of Cyprus—both produced by the dynamics of colliding plates and subducting crust.

Though agriculture arose in this region about 9,000 years ago and ceramics had developed there by 8,000 years ago, it was not until about 6,000 years ago that the first copper objects become a significant part of the archaeological record. A well-made flint ax may have been sharper than one of copper, but the stone is brittle and must be thick if it is to be kept from breaking. The copper ax could be narrower and was more durable. At first, metals were used to make ornamental or religious objects as well as the tools of war; only later did metal become widely available for more ordinary uses.

Working with metals required several labor-intensive and coordinated stages of activities, beginning with mining the ore. Pure metals, in iron meteorites or in rare veins of pure copper, had been used earlier but were quickly exhausted. Mining of ores, based on the same techniques used earlier for quarrying flint, became a serious enterprise as the demand for metals escalated. Once the ore was available, it had to be refined by separating the metal from the oxygen, sulfur, and other elements with which it was combined. This could be done only at high temperatures in a furnace to carry on the necessary reactions, to remove impurities, and to separate the slag. The earliest evidence of copper smelting is dated at about 7,000 years ago in the Middle East. By 5,500 years ago, metalworkers had learned that the addition of tin or arsenic to copper could produce bronze, a harder metal with greater tensile strength. Bronze also has a lower melting point, making it easier to cast into molds to produce a wide variety of useful and ornamental objects.

It may be that in the Middle East the separation of copper from its ore was discovered by chance, when malachite was accidently introduced while firing ceramics. A copper-containing mineral common in some parts of the Mediterranean and a favorite material for jewelry, malachite could have accidently fallen into the hearth, producing the precious metal by accident. The mastery of iron was much more difficult, but it, too, may have been the product of fortuitous circumstances.[29] Having copper and bronze technology as a background and experimenting with a variety of materials, early metalworkers did not master the use of iron until about 3,500 years ago, *again* in the Middle East. "Serendipitously, the most common flux used for copper ores [in the Middle East] was hematite, which, we now know, is an iron oxide. The carbon dioxide–carbon monoxide gas mixture from the reaction of charcoal with oxygen during copper smelting can be exactly what is needed to reduce hematite to metallic iron," notes materials scientist Stephen Sass.[30] Finding small pieces of metallic iron while producing copper may have been the accidental but necessary first step in learning how to extract iron from its ore. Here's another grand accident in the history of *Homo sapiens*; it seems unlikely that people would have discovered that hematite contains iron in any other way.

Using trial and error, early metalworkers elaborated their methods with absolutely no understanding of the chemistry of these reactions; more than two millennia would pass before materials scientists finally understood these processes. And, just as a small amount of arsenic or tin could make copper a far more useful metal, a small amount of carbon would transform iron into the metal we now call steel. Because their furnaces could not reach a high enough temperature to liquify pure iron, carbon was added slowly through diffusion by heating iron for a longer time in a charcoal fire. Early workers actually thought they were purifying the iron during this process of "carburization." Once hot-forged iron had been carburized to just the right composition, the tempering by hot-hammering and quenching allowed the ironsmith to produce a metal with twice the strength of bronze. A well-made bronze sword may have been superior to an early iron version, but once more efficient techniques were developed, iron could be produced much more cheaply than bronze.[31] Around 2000 B.C.E. four grams of silver could purchase only one gram of iron, but by 700 B.C.E., one gram of silver would purchase two thousand grams of iron. With lowered costs, working tools of iron soon joined the murderous artifacts of warfare in the archaeological record—and another momentous step in the human saga had been achieved. Originating in the Middle East, iron technology soon spread to Africa, southern Asia, and China.

The rise of metallurgy in the Middle East was not merely a triumph of human ingenuity and perseverance, it was also the good luck of having large, settled, stable agricultural societies in close proximity to a rich geological environment. The first major step had taken place about 8,000 years ago with the use of clay to make hot-fired ceramic pottery in kilns. The first large cities of western Asia arose about 6,000 years ago, followed by the Bronze Age about 1,000 years later. Though early American cultures did create exquisite gold work and some copper and bronze artifacts, the peoples of the Americas produced very few metalworking tools. The architectural achievements of the Mayas and Incas are all the more remarkable when one remembers that all their monuments were built without the use of metal tools.

While copper production developed independently in several areas, the efficient production of iron and steel appears to have originated uniquely in the Middle East, and this important technology was transferred throughout the Old World from that solitary origin. Also, it seems unlikely that early societies could have learned to extract iron unless more easily worked metals such as copper, lead, and tin were already available in concentrated ores thanks to dynamic tectonic processes. The development of metallurgy is yet another fateful historical contingency in our long epic, thanks to metal-rich rocks readily accessible to major agricultural civilizations.

## A GROWING THIRST FOR POWER

With agriculture and metallurgy, early civilizations had what they needed to claim dominion over much of their surroundings. There were, however, several other important steps before we humans became the masters of our environment that we are today. One of the most basic was achieving access to more "power." Cultivated plants and domestic animals were the first major steps, dramatically increasing the food energy and draught power available to human communities. While an ancient farmer could feed himself and, in good times, a dozen other people, the modern mechanized farmer can feed thousands. Contrary to popular assumptions, a modern farmer is not inherently more efficient than the earliest farmers of the Nile or Yangtze flood plains. What the modern farmer has available is power, as expressed by large tractors, concentrated fertilizers, herbicides, insecticides, huge combines, and sophisticated drying, storage, and shipping facilities. A modern farmer's ability to produce and store a huge harvest is no longer based on human or animal labor but on the energy of fossil fuels powering the internal combustion engines that keep his tractors going. Likewise, modern fertilizers, insecticides, and drying equipment are directly or indirectly linked to the spinning turbines at regional electric generating plants, in turn powered by fossil fuels, moving water, or atomic chain reactions.

The ancient Romans had constructed waterwheels as a potential source of power, but animals and cheap slave labor resulted in little technological development during the Roman period. After the fall of Rome, western European history entered a period that is often called the Dark Ages, but they weren't particularly dark for long. Soon, the late Middle Ages would become "one of the great inventive eras of mankind"—beginning with important innovations in agriculture.[32] The moldboard plow, capable of tilling heavier northern soils, was introduced to western Europe after the sixth century. A heavier wheeled plow was developed around the eleventh century; plowing deeper, it allowed harrowing to replace the work of cross-plowing. Nailed horseshoes, cushioned horse collars, and the breeding of stronger horses allowed European farmers to plow faster and longer than with oxen. Oats and newly introduced alfalfa helped feed the hard-working horses. All these factors, together with better crop rotation, fed a rapidly growing population. An especially important role was played by carefully tended gardens on estates and at monasteries, serving as small experimental plots in which new introductions could be monitored and selected.

Based on the prices for wheat and bread, historians have inferred that the climate of Europe was especially favorable between about 900 and 1314 C.E.—and the region prospered.[33] Population increases resulted in an explosive growth in the number of towns, expansion of trade, and a demand for more goods and services. All these trends provided a strong incentive to expand the use of water power. At Rouen, along a minor tributary of the Seine River in central France, there were two water mills in the tenth century, five in the twelfth, and ten in the thirteenth. The Domesday book, surveying thirty-four English counties, listed 5,624 water mills in 1086. At first, mills were used to grind grain into flour, but they were soon modified to saw wood, crush rocks, and provide power in tanning and cloth fulling; later they were used in paper making and to work the bellows in blast furnaces. Benefiting from generous year-round rainfall and many continously flowing streams, vertical waterwheels provided Europeans with a generous source of power.

Windmills, a technology developed in western Asia, were transformed in western Europe. With their shafts held horizontal and rotatable to face the wind, windmills helped the Netherlands drain large areas of marshland. Both water mills and windmills were major capital expenditures; keeping them repaired and profitable required a busy mercantile environment. With a growing market economy based on active local and international trade, using freely exchangeable currencies and bills of exchange, Europe of the later Middle Ages moved boldly forward in what has been called an "economic revolution."[34]

Europe increased its productivity in many different ways. Women and children were an important part of the economy; contributing many long hours in gardening, food preparation, and the production of textiles. A reverence for work and the threat of starvation helped justify so much effort. Minor innovations also increased productivity; eyeglasses doubled the working life of artisans and scribes in the early fourteenth century. The mechanical clock divided time into easily counted fragments through the use of oscillatory motion. Early medieval clocks were large and communal, ringing bells in the public square. Smaller clocks soon took their place in the workplace; they promoted punctuality. "It is the mechanical clock that made possible, for better or worse, a civilization attentive to the passage of time, hence to productivity and performance," declares David Landes.[35] A respect for timeliness helped Europe deal with ever more complex systems of mercantile interchange. Accurate timepieces were especially important for ocean navigation. Finally, clock makers became the precision metalworkers who would construct the instruments and machinery of science and industry.

Despite the decimation caused by the black death (beginning in 1347), Europe's mercantile activities soon resumed at full gallop. Spurred on by warfare and exploration, Europe's thirst for energy continued to escalate. Year-round rivers for water power, the availability of timber and coal, and the presence of iron mines had shifted the center of European civilization from the Mediterranean toward central and western Europe. But as the centuries passed, the use of wood in building warships, as fuel for iron smelting, brick making, and glass making, together with the more mundane needs of cooking, heating, and construction soon decimated forest and woodland. As timber became scarce and costly, increasing demand for coal as a heating agent presented mining officials with a serious problem: how to remove water from flooding coal mines. The first steam engine was developed for this purpose around 1712 in England, but it was not until the invention of a more efficient steam engine by James Watt in the 1780s that steam power became an effective way of driving machinery for a wide variety of operations.

The steam engine converted the heat of burning wood and coal into rotary power, transferable through gears, chains, and belts for many kinds of enterprises. Better yet, the steam engine could be placed where it was needed, unlike water mills or windmills. The British Isles, less troubled by warfare than the nations of continental Europe, and with both coal and iron ore in abundance within its borders, pioneered the production of steel machinery. An efficient banking system, both public and private, made possible the funding of capital-intensive enterprises. All these concurrent developments gave rise to yet another major advance in our odyssey: the industrial revolution. Powered machinery could spin cotton into threads two hundred times faster than human hands. Cheaper cotton clothing was much easier to wash than wool or leather, and together with other advances in sanitation, contributed significantly to improving health. Factories of complex powered machinery were now transforming raw material into mass-market merchandise for a population growing in both numbers and purchasing power. Whereas Britain had only one large city and used about 11 million tons of coal a year in 1800, by the 1890s Britain had twenty large cities and used over 100 million tons of coal each year. Here was where the industrial revolution began.

Sequestered in the universities, and despite being part of a shared culture in which problem-solving activities were highly regarded, academic science contributed little to the origins of the industrial revolution. The induction of an electric current within a copper wire by a spinning magnet was discovered in academia, but transforming that insight into an electric motor and the long-distance transmission of electric power was accomplished outside of the university community. Not until the late nineteenth century were the elements of the scientific and industrial traditions effectively joined as a powerful strategy for accelerating technological progress. By fusing the latest scientific research protocols with specific industrial objectives in laboratories receiving government support, German corporations set the pattern for further rapid innovation. Having united science and technology, the industrial revolution continues to transform our world.[36]

Between 1810 and 1910, the industrial revolution created inventions of profound consequence: rapid rail transport, innovative agricultural machinery, more efficient steel making, precisely fashioned interchangeable parts and mass production, electric generation and long-distance power transmission, the telegraph and telephone, photography, the light bulb and phonograph, the internal combustion engine and automotive transport, industrial chemistry, mechanical refrigeration, wireless telegraphy, and heavier-than-air flight. Within less than a hundred years, and having begun with a gliderlike biplane constructed by two bicycle mechanics, we've flung spacecraft beyond the solar system.[37] The alacrity of recent technological progress is unparalled in human history—and tightly linked to our ravenous appetite for energy. J. R. McNeill estimates that humankind increased its energy consumption fivefold in the nineteenth century—and another sixteenfold in the twentieth.[38]

The industrial revolution, like the agricultural revolution that preceded it, changed the world. Many factors played a role in this new transformation, not the least of which was the introduction of new food crops from the Americas. Brought from the highlands of the Andes, potatoes produced four times as many calories per acre as did rye in the cool moist summers of northern Europe. "American food crops were an essential resource for the nineteenth century surge of numbers, wealth, and power that raised European nations so far above the rest of the world," claims William McNeill.[39] This increase of available food resources helped power the Western juggernaut. The explosion of human numbers has since been further accelerated by what Lindsey Grant claims, perhaps correctly, was the greatest invention of the early twentieth century: the Haber-Bosch process.[40] Using high temperatures and special catalysts, this process unites gaseous hydrogen and nitrogen to form ammonia, the basis for making nitrogenous agricultural fertilizers. In turn, these fertilizers are now essential in feeding a world of six billion people.

With the development of the internal combustion engine in the 1890s, fossil oil deposits joined coal and water power as a major source of energy; our appetite for clean-burning natural gas is a more recent phenomenon. Meanwhile, electric generation stations and high voltage transmission lines have made power accessible to home, office, and factory. In the industrial world, muscle power has been largely replaced by electric motors and diesel engines, allowing for a phenomenal increase in what we call "productivity" (actually it's mostly more fossil fuel consumption). The computer-powered information revolution of today is only the latest episode of these cumulative changes. Our insatiable consumption of energy has made each of us far richer than King Solomon, watching digitized movies on our wide-screen surround-sound home theaters and, for a very few, listening for radio messages from outer space.

Our increasing consumption of energy has played a critical role in our progressive cultural advancements over the last ten thousand years. We can view this escalating use of energy from the perspective of Van Valen's Red Queen; we have seized an increasing share of our planet's available energy. Both the agricultural and industrial revolutions have been powerful progressive advances in the "fitness" of the human species. By tilling the soil and herding grazing animals, early agriculturalists transformed natural ecosystems, boldly expanding our share of the solar energy shining down upon us. Then we discovered the sequestered energy of millions of years of photosynthetic activity in fossil fuels, thanks to long-term processes of geological accumulation within the planet's crust. Today we're sucking up fossil oil at a rate somewhere around 75 million barrels *each day*, and this doesn't include coal, gas, or peat. Hydroelectric, wind, geothermal, and nuclear power have also contributed to our insatiable appetite. These sources of power have made possible changes in our lives that could not even have been imagined in earlier times. However, neither modern industrial capabilities, our recent increases in energy consumption, nor building radio telescopes is likely to have been possible without an earlier and more fundamental innovation in human culture; today we call that transformation the scientific revolution.

# 10

# THE ORIGINS OF WESTERN SCIENCE

clever species

Modern science has transformed our lives and our understanding of the cosmos. A continuing flood of amazing new discoveries and complex new inventions gives the impression that our scientific society is part of a progressive trajectory that is just about inevitable. This same intellectual environment makes one think that intelligent beings circling other stars can't help but follow a similar path. Modern science seems to be the natural outcome of a progressively richer and more elaborate historical journey. In Western history, the civilizations of Sumeria and Egypt were soon followed by the incisive intellectual curiosity of classical Greece. An expansive Roman empire was followed by a period of decline in the West, but soon eclipsed by Renaissance and the birth of modern science. In such a broad historical context, progress may seem inevitable. However, a closer look at what happened during medieval times will, I believe, give us a very different view. The origin of Western science was not the simple product of predestined historical trends.

A series of events came together felicitously in Western European society beginning in the tenth century. To start with, the climate improved, just as the threat of barbarian invasions diminished. Good weather continued from the tenth to the early fourteenth century. With improved farming practices, rising populations, expanding trade, and widely exchangeable currencies, medieval Europe began a remarkable transformation. The number of towns increased *tenfold* between 900 and 1350 C.E.[1] These towns supported a growing class of shopkeepers, traders, and craftsmen, complementing the feudal landowners and peasant farmers. Often developing next to fortified castles along the junctions of rivers or trading routes, towns and their populations expanded. Soon these communities would declare their prosperity and importance by building larger and more graceful churches with walls made luminous by stained glass windows. These buildings, public in every respect, were the vital spiritual and ceremonial center of the community—and the region's most important source of civic pride. In France alone, more stone was quarried in this period than in all 3,000 years of ancient Egypt's history.[2] These buildings are a bold reflection of a more vibrant commercial society.

The gothic cathedrals of western Europe are one of the finest architectural achievements in all of human history. During the eleventh century, architects had begun to replace fire-prone wooden roofs with a barrel-vaulted ceiling of stone. Inspired by the success of these late Romanesque churches, a rage to build taller edifices swept across western Europe. As their expertise improved, master builders were able to create thin but carefully buttressed walls of extraordinary height, capped by high ceilings of cross-ribbed vaults with pointed arches.

One of the earliest of the great French cathedrals, Notre Dame de Chartres, stands tall above its community in the midst of gently undulating wheat fields about fifty miles southwest of Paris. An important destination for pilgrims, the cathedral had burned in 1030 and burned again on 10 June 1194.[3] A major effort was immediately made to rebuild the monument. The visiting worshippers and penitents had an important impact on the economic well-being of this town. Immediately after the fire, appeals went far and wide to raise money for reconstruction. Quickly, the new building began to rise. Considering the scale of houses within the community at the time, this new church was impressive. From west door to altar, the central aisle measures 427 feet long; the floor plan covers 65,000 square feet, and the ceiling above the altar soars to 123 feet. Thanks to sturdy flying buttresses, helping deflect the outward thrust of a heavy roof, the slender piers frame more than 150 windows. Four thousand sculptures decorate the cathedral's exterior, repeating many of the themes visible in the stained glass from within. Portals at north and south transepts and the west entrance stand beneath three large circular rose windows. Emblazoned with stained glass of deep hues that have never been surpassed, the radially patterned rose windows illuminate the entries. Thirty-three feet in diameter, these windows were especially costly to construct. Between 1194 and the cathedral's consecration in 1260, this small town had created one of humankind's most wondrous architectural achievements.

More central to the elaboration of our theme than the rebuilding of Chartres cathedral itself was its associated school. Charlemagne, Holy Roman emperor who ruled from 768 to 814, had mandated that all cathedrals and monasteries establish schools to educate the clergy, and these became the institutions for learning during a time when cities were in decline and barbarian invasions were still a threat. An educational center for several centuries, the school of Chartres gained renown in the middle of the twelfth century for a small but influential faculty who taught and discussed faith and philosophy. It was here that William of Conches's teaching prompted charges of heresy. One of the ideas that got William in trouble was the notion of nature acting as an autonomous rational entity. "William's purpose," writes David Lindberg, "was not to deny divine agency but to declare that God customarily works through natural powers and the philosopher's task is to push those powers to their explanatory limits."[4]

These scholarly activities had been impelled forward by a unique historical coincidence. With the retreat of the Muslims from Toledo, Spain, a major Islamic library became accessible to European scholars in 1085. Additional Muslim libraries were brought to Toledo, and a center for translation was established there. Together with the work of scholars in Sicily and Greece, the core of Islamic science and earlier Greek science became available to Latin-reading western Europe. The Greek achievement had been lost to western Europe after Christian civilization became divided into western Latin-speaking and eastern Greek-speaking halves. Suddenly, over a thousand years of Greek knowledge regarding astronomy, mathematics, medicine, and natural history, preserved and expanded by Arabic scholars, was made accessible to readers of Latin. The writings of Islamic and Hebrew philosophers were also newly revealed through new translations. In the words of Richard Tarnas: "Medieval Europe's sudden encounter with a sophisticated scientific cosmology, encyclopedic in breadth and intricately coherent, was dazzling to a culture that had been largely ignorant of these writings and ideas for centuries."[5]

The great cathedral itself bears witness to these scholarly activities. Although the edifice offered the community a superb monument to its belief, it also reflected a profound respect for knowledge of the real world. Small carved figures of the mathematicians Pythagoras and Euclid, Ptolemy the astronomer, and Aristotle the natural philosopher decorate one of the west portals. Here at Chartres, remarks Norman Cantor, is where "the great intellectual quests of western civilization resumed."[6] (However, it was by no means the first, and many other schools, such as Paris and Oxford, played important roles as well.)

Aristotle's texts, enriched by Arabic commentary, claimed that the world could be explained on the basis of fundamental elements, causal properties, and rational inquiry. Christian philosophers, working within a cosmology governed by a single God, were able to unite the celestial and earthly realms that Aristotle had viewed as separate. According to Edward Grant, this new philosophical orientation claimed that "God had conferred on nature the power and ability to cause things. . . . Nature, or the Cosmos, was thus objectified and conceived as a harmonious, lawful, well-ordered, self-sufficient whole which could be investigated by the human intellect."[7] Judaism's solitary deity had been the foundation of these attitudes. "A single, exclusive, jealous and iconoclastic deity, averse to magic and graven images was perhaps the most decisive formative influence in the education of the human race," claims Ernst Gellner.[8] From such a foundation, natural philosophers, having promoted the idea that nature operated according to inherent laws that were a significant aspect of God's work, inspired the serious study of nature as a means of *knowing* God. In addition, there were many important Islamic, Hebrew, agnostic, and atheistic natural philosophers who contributed greatly to the intellectual environment from which science would grow.

## CADAVERS AND THE ORIGINS OF WESTERN SCIENCE

Modern Western medicine is rooted in the early work of Hippocrates and his associates, who avoided magical, mythical or theological approaches to the curing of disease, focusing rather on repetitive observation and analysis. By the third century B.C.E. the Greeks had solidified this tradition and developed an impressive herbal pharmacology. Galen added a large body of medical work in second-century Rome; but as Rome declined, much of this knowledge was salvaged by scholars and medical practitioners in Greek-speaking Byzantium and the Muslim world. Translated into Arabic, the earlier medical knowledge was further advanced as Islamic culture blossomed.

King Frederick II, who founded the University of Naples in 1224, may have been the first to have promoted dissection of human cadavers in Europe. His treatise on falconry in 1245, *On the Art of Hunting with Birds*, included original descriptions based on his own dissections of the birds he had studied. By the end of the thirteenth century, several medical faculties in northern Italy began regular postmortem examinations. In 1380, perhaps as a consequence of the great plague, medical schools in Italy were given permission to use the bodies of the executed and unclaimed for use in dissection. They were limited to only a few bodies each year, and confined this activity to the coldest time of winter for obvious reasons (formaldehyde and refrigeration were far in the future). The dissections were teaching demonstrations affording firsthand observation, but lectures were still based entirely on ancient texts. Italian medical faculties attracted scholars from all over Europe to study what had never been studied so intensively before: the human body.[9]

Human societies have always had specific and restrictive conventions regarding how the bodies of the deceased are dealt with. These social prohibitions remind us that cutting up the newly dead and examining their insides is a rather bizarre activity, far beyond the pale of ordinary behavior and outlawed in almost every culture of the world. Historians of science have focused primarily on the astronomers, physicists, and chemists of the sixteenth and seventeenth centuries when discussing early Western science. However, I believe that those pursuing the investigation of human anatomy give us a better insight into the unusual nature of early Western science. Scholars and alchemists had contemplated the stars and the nature of matter in many earlier societies, but how many were cutting up dead people, and getting respected artists to join in this smelly enterprise? Something new and different was happening in European society.

Galen, who lived from 129 to ca. 200 C.E., based many of his medical texts on the dissection of barbary apes and pigs since the Roman government prohibited the dissection of human bodies. Keep in mind that this was the same government providing bloody human combat as a form of entertainment in the Coliseum and allowing the substitution of convicted felons for actors in onstage executions. The critical point here is simple: no previous society had created durable medical institutions in which the careful study of publicly dissected human cadavers was sanctioned.[10] Nor was this interest in studying nature directly limited to the medical professionals. In making his own detailed studies of human anatomy in the late 1400s, Leonardo da Vinci dissected at least fourteen cadavers before being barred by local authorities from further access to the mortuary. Having one of the world's greatest artists also caught up in this malodorous enterprise is, I think, profoundly significant. Neither a physician nor a medical student, Leonardo insisted on studying the human body himself. His superb anatomical illustrations, and most of his other studies, were never widely circulated, with the result that he contributed little to the growth of science, but the fact that such a gifted intellect would pursue these many inquiries in such intense fashion is important. For many, the "book of nature" had become a new path for achieving knowledge of the divine. For others, however, the appeal of science was more practical and earthbound; they were trying to change lead into gold or discover new cures.

Fundamentally, science is a sociological phenomenon. People usually do what they think will impress others. Prestige and honor are second only to power and money. For however many reasons, it became *fashionable* for men in the upper strata of society in western Europe to investigate nature. They studied the firmament, analyzed the forces of nature, illustrated medicinal herbs, and dissected cadavers. As Roy Porter points out, "Dissection was justified largely in terms of natural philosophy and piety (the body demonstrated the wisdom of the Creator); the surgical benefits were rarely mentioned. . . ."[11] And though they represented a higher social stratum that could afford to spend time in such eccentric studies, these scholars were not isolated from the craftsmen who might fashion their instruments or engrave their illustrations. Nor did these "natural philosophers" lose status in the community because they had soiled themselves with their efforts; the intense study of nature was now generally acknowledged as a worthy effort in comprehending the majesty of the cosmos.

As Europe flourished, its population continued to rise, resulting in more crowding and poverty, less sanitary conditions, and recurrent minor epidemics. Burgeoning trade produced a dynamic prosperity, but it also brought danger. Dispersed by rats and fleas, a virulent pathogen arrived from the East; beginning in 1347, no area of Europe escaped the black death. Bubonic plague and its accomplice, pneumonic plague, oppressed Europe with recurring outbreaks for almost twenty years. Overall it is estimated that somewhere between 30 and 40 percent of the European population perished during this period. Nevertheless, mercantile activities recovered quickly. The population crash caused the value of labor to increase dramatically and the price of land to drop, resulting in a rise in the standard of living. Here's another instance in our long history where a calamity gave rise to a positive outcome. The black death may have been an important prelude in the flowering of the Renaissance.

Continuing economic prosperity of northern Italy's city-states had renewed an interest in the study of the classics, which were understood to represent a different culture. As scholars became aware of civilizations long past, the possibility of historical progress and the growth of knowledge itself became manifest. At first, rediscovery of ancient wisdom was the motivating force, but soon new discoveries from around the world opened new vistas. Exploration for treasure or trade would itself bring in a flood of new information and challenge the validity of ancient knowledge.

## VOYAGES OF DISCOVERY

Perhaps we can think of the late-medieval mariners as the first practitioners of that art we now call science. Studying charts carefully prepared from earlier voyages, they made hypotheses regarding further uncharted shores, then set sail to test those conjectures. Columbus, both a sailor and chart maker, and using his considerable navigation experience, sailed south toward the Canaries to catch the easterly winds, then turned his ships westward into the unknown. He had a bold hypothesis and was ready to risk his life for its validation.

Introduction of the magnetic compass from China in the late 1200s had provided European mariners with a critical new aid in navigation. Combining both Mediterranean and Atlantic designs, shipbuilders developed a strong new three-masted ship, steered by a single sternpost rudder. Smaller sails, both square-rigged and lateen, allowed the ship to tack more effectively against the wind. Tough little ships that could be sailed by a small crew, the caravels were built to explore the unknown and bring their crews home again. Similar in design but built to bring back precious merchandise, larger carracks could haul well over a hundred tons. With overland trade from Asia in the iron grip of Venetian traders and Muslim middlemen, together with the hazards of overland travel through Asia and the Middle East, prices remained high. Any ship that could circumvent this bottleneck loaded with spices, silks, and ceramics would make a fortune.

After Muslim sovereignty had been removed from the Iberian peninsula, Portugal began an unprecedented program of intensely focused exploration. In 1420, Prince Henry of Portugal (1394–1460) founded a center for the study of navigation with a library for the storage of maps and manuscripts. This center included an observatory and supported mathematicians and astronomers; here, simple navigational instruments and trigonometric tables were created by which ship's captains could estimate their latitude along the African coast. Henry's support had secured the Canaries, the Azores, and the Madiera islands as supply stations for an extraordinary effort, the circumnavigation of Africa. This was Europe's first carefully organized long-term program of ocean exploration for trade.[12] Moving ever farther along the coast of Africa, Portuguese ships crossed the equator in 1471; they reached the Congo River in 1483, and rounded the Cape of Good Hope in 1486. In that same year, King John II, a successor to Henry, sent an Arabic-speaking emissary via Aden (in Yemen) to learn the routes that Muslim ships were using in the spice trade. Sending valuable information back to Portugal, Pero da Covilhá had reached Calicut on the western shore of India. Here, ships offloaded precious cargoes of spices, silks, porcelains, and jewelry from the east. On the basis of this information, and with the knowledge acquired by earlier voyages, Vasco da Gama set forth with three ships and a two-hundred-ton supply vessel on 8 July 1497. Their goal was to reach India. To avoid contrary winds along Africa's long coastline and take advantage of the southern westerlies, the expedition sailed southwestward toward Brazil, before turning eastward for Africa's cape. Sailing almost 4,000 miles for 137 days, and knowing the precise latitude of Africa's southernmost point, they made their first landfall after rounding the cape. Here the supply vessel was broken up and the three ships reprovisioned. Sailing northward along the eastern shore of Africa, the expedition stopped at Mozambique and Mombasa, active ports in centuries-old trade between eastern Africa and Asia. At Malindi, Kenya, the Portuguese expedition had the extraordinary good fortune to hire an expert pilot, Ibn Majid, who guided them across the Indian Ocean to Calicut. Returning to Portugal after a journey of two years and 24,000 nautical miles, the original crew of around 170 had been reduced to two ships and only 55 men (poor nutrition had taken the majority). One of the most ambitious achievements in the annals of discovery, the expedition had reached India by sailing around Africa, returning with a cargo of spices commandeered from an Arab vessel.[13]

Using the brute force of superior military and naval technology, the Portuguese quickly came to dominate trading routes along the south Asian coast. When confronted by much larger naval fleets whose only means of combat was ramming and boarding, Portuguese ships kept their distance and let loose with cannon fire; they had cannon balls to spare. Portugal's program to circumnavigate Africa had been inspired by capitalism's driving force: money to be made. The great costs of the first exploratory voyages, possible only with the Crown's backing during a century of political stability, proved to be an astute investment. In 1503 alone, Portuguese shipping brought 1,300 tons of black pepper around Africa to Europe; soon the price for pepper in Lisbon fell to a fifth of the price in Venice. For more than a hundred years, Portugal provided Europe with Asian imports at the lowest prices.

Much earlier, a long history of Mediterranean maritime commerce had been further energized when state-guaranteed coinage was introduced in Anatolia, Turkey, around 680 B.C.E. Incessant wars and continuing political upheavals notwithstanding, commercial shipping persisted over the centuries. The Atlantic Ocean, once the impenetrable edge of the Western world, now became the threshold for launching an assault on the entire planet. With their own long history of continual warfare, multi-directional internal migrations, active commerce lubricated by effective financial instruments, and an evangelical religion, Europeans launched an expansionary imperialism that would change the world and themselves.[14]

Late medieval voyages of commerce and discovery produced important ancillary effects for the development of science. Eyewitness accounts by returning Portuguese sailors contradicted Aristotle and Ptolemy: the tropics were neither torrid deserts nor uninhabitable. Such a clear refutation of thousand-year-old textual dogma was no small matter in a culture still inclined to accept ancient wisdom as infallible. Here was further proof that the validation of knowledge required new and continuing investigation. With circumnavigation came verification for the claim that we live on a sphere surrounded by a seamless vault of stars and, more important, that the laws of nature are everywhere the same. As John Hale phrases it: "Theory could only be verified in the field of discovery through the cooperation between the hypothesizers at home and the observers in their ships. The first scientific laboratory was the world itself."[15]

## PAPER AND PRINTING

Until the fifth century, Western texts had been written and preserved mostly as papyrus rolls; thereafter bound books became the standard format with the use of parchment, derived from the skins of sheep or calves. Once folded, a single sheepskin became the folio; further divisions produced the quarto and octavo sizes. A single two-hundred-page volume might take a scribe more than four months to copy and require twenty-five sheepskins; these were costs only the wealthy or major ecclesiastic centers could afford. Scholarship was possible only in close association with a major center of learning or a royal patron. Few libraries exceeded three hundred volumes and many secured their books by chains just long enough to reach a nearby reading table. Books were exceedingly precious, closely guarded, and difficult to access.

Created from the inner bark of the paper mulberry and hemp fibers, paper was first produced in early second-century China. With the capture of Samarkand in 751 C.E., a Chinese paper-making mill became known to the Muslim world and such facilities were soon built in Baghdad and Damascus. However, the first paper mills were not established in Italy until 1268. Parchment was expensive and papyrus had been fragile, but the new paper based on linen and other plant fibers was strong, flexible, and reasonably economical. Woodblock printed religious verses, figures of saints, and playing cards, all made of paper, became popular and profitable in Europe by the early 1400s.

Moveable baked-clay letter fonts had first been made in China, but these were fragile. Metal typecasting had been developed in Korea by 1313. Printed by the government and distributed without cost, books and book making did not expand further in Korea. In the busy mercantile environment of Europe, however, one ambitious craftsman and his associates recognized the commercial potential of the new technology. Using his considerable metalworking skills, trained assistants, and modifications of earlier printing techniques, Johann Gutenberg in Mainz, Germany, created an explosive advance in the communication of knowledge.

A print shop was no small enterprise. Starting with the construction of finely hand-crafted hard-metal punches, the letters were impressed in a soft brass matrix, which was then placed in a mold. Hot molten-lead alloy would then be poured into the mold to produce the individual type elements. This technique, using the same three-parted mold, was capable of quickly producing many precisely interchangeable pieces of metal type. Even with only twenty-six letters in the Latin alphabet, each size of type would require at least 150 different elements. And that was just the beginning. Gutenberg's shop probably employed a dozen people to create and repair the metal type, set and block the type, print the proofs, read and correct the proofs, print the corrected pages, then dry and stack the pages. Having been a skilled goldsmith, Gutenberg was able to develop a technique for precisely casting metallic hard type that would stand up to repetitive printing. Gutenberg's other innovations included insuring proper alignment of lines of type, adapting and improving the screw-and-lever press, and designing means for holding the paper precisely in place. Because of variations in paper thickness and height of type, Gutenberg used oil-based ink on moistened paper.

After many years of effort, the Mainz workshop (now under the control of his creditors) produced the first major printed work in 1455, a 1,282-page Bible. Each forty-two-line, double-columned folio page of that work contains about 2,750 letters. Michael Clapham estimated that printing six pages simultaneously would have required as many as 100,000 metal type elements.[16] With skilled craftsmen and significant capital investment, this new enterprise made possible the production of inexpensive but highly profitable books.

A superb example of technological innovation in the context of what we now call venture capitalism, printers quickly spread from Germany to set up shop in all the major cities of Europe. By the end of the fifteenth century, printing with movable type had produced over 35,000 titles and editions, numbering an estimated five million volumes. No longer laboriously produced by hand copying, books became a readily available source of information and propaganda, as well as highly desirable merchandise. Printing with movable type transformed the communication of knowledge. Earlier, in order to build a new library quickly, Cosimo de' Medici (1389–1464) had contracted to have fifty-five scribes copy books; they produced two hundred volumes in twenty-two months. Compare that effort with the situation in 1516, when a single print shop produced thousands of copies of Erasmus's *Colloquies* in a few months. With Gutenberg's new printing technology, western Europe was able to create a larger number of books within a century than had been produced on the entire planet over all previous time. From a technological perspective, this achievement can be epitomized in a single statistic: Gutenberg's technique was not substantially improved for another three hundred years.

Thanks to this powerful invention and his own vernacular translations, Martin Luther (1483–1546) could insist that all literate believers study the Bible for themselves. Protestanism's scripturalism and reader-centered theology, in turn, helped develop an individualistic approach to the acquisition of knowledge. In such an environment, scientific investigators could do research in relative isolation and communicate their findings through printed journals to a widely dispersed community of scholars. By helping bring about the Reformation and accelerating the dissemination of knowledge, the printing press played an essential role in catapulting the scientific enterprise forward.

## THE SCIENTIFIC REVOLUTION

The scientific revolution was no sudden overturning of previous thinking. Science is neither a unified body of knowledge nor a monolithic enterprise; different disciplines have progressed at very different rates. Copernicus's revolutionary view of a heliocentric solar system claimed that the Earth rotated once each day, revolved about the Sun once each year, and produced the seasons because of a tilted axis. Using the careful observations of Tycho Brahe (1546–1601), and after having seen recent translations of Greek texts on conic sections, Johannes Kepler (1571–1630) realized that the planets did not move in perfect circles as Copernicus had thought. Instead, they were moving in slightly elliptic orbits and at varying speeds, sweeping out equal areas of the orbital plane over equal time. Using Kepler's insights and his own mathematical innovations, Issac Newton (1642–1727) was able to explain the gravitational nature of planetary motion only 144 years after the publication of Copernicus's innovative text. Newton's magnificent synthesis marked the epiphany of the "Copernican revolution." However, other areas of science, such as medicine, progressed far more slowly. It was not until three hundred years after Vesalius published his superb treatise on human anatomy that the nature of disease would be subjected to serious scientific analysis.

Slowly the religious conviction of a fixed higher truth, determined by divine scripture and "logic," began to be eroded by a torrent of new information. William of Ockham had argued for separating religious revelation from the work of reason and science. Even earlier, Robert Grosseteste and Roger Bacon had emphasized the importance of observation and experiment. But it was not until three hundred years later that experimental science would be epitomized in the efforts of Galileo Galilei (1564–1642), whose demonstrations, inventions, discoveries, and confrontational popular essays shifted the scientific enterprise into high gear. His accomplishment is especially impressive when we consider the environment of persecution and intimidation that existed at that time. Near the end of his life, Galileo was convicted of heresy by the Catholic Church. Giordano Bruno was burned at the stake. Accusations of witchcraft were commonplace and religious intolerance was widespread.

Despite such a hostile atmosphere, the scientific enterprise continued. Another Italian, Francesco Redi, performed a series of experiments in 1668 to demonstrate that maggots would not develop in aging meat if flies were excluded. Experimentation had now become the preferred technique for both confirmation and falsification of hypotheses. It was perhaps no accident that science developed in those countries where the making of instruments central to the measurement and repeatability of many experiments was most advanced. In this new environment, "Whatever did not lend itself to the modern methods of verification and scrutiny was banished from the realms of science and branded with the stigma of fabrication or idle fantasy."[17] With the invention of new instruments, the experiments of Robert Boyle, the writings of Descartes, and the mathematical majesty of Newton's laws of gravitation, the scientific revolution developed a momentum that could no longer be contained. The methodology of Western science had now become established as a major progressive force in the story of our species. As Richard Powers phrases it: "No single idea has had a more profound or ubiquitous impact on what the human race has become, or what it has worked upon the face of the planet, than the vesting of authority in experiment."[18]

While some scientific studies had been inspired by a deeply religious purpose in the early medieval period, as the centuries passed science reached ever further into the historical depths of nature and would challenge the creation story of Genesis. Geologists' continuing study of rock formations made it clear that our planet was much older than anyone had imagined. Lamarck and others interpreted the patterns of diversity among living things to be the result of a long history of evolutionary change. Darwin and Wallace explained these many evolutionary trajectories by proposing the commonsense notion of "natural selection" as sufficient for creating the manifest perfection we find in nature and making that process amenable to scientific inquiry. More subtle biological and geological phenomena revealed their secrets only after many further decades of intense effort. The molecular architecture of the genetic code was not deciphered until 1953. With new imaging techniques, neurologists are now seeing in graphic imagery how mind and brain function in real time. Progress in science and technology has been explosive over the last two hundred years and continues to grow at an almost exponential rate. It seems incredible that we are using giant radio telescopes to probe the universe today, when the existence of radio waves was completely unknown before 1887. But why did this progressive trajectory gain momentum in Europe?

There are times and places where new and important findings cannot find an appropriate environment in which to take root. Mendel's demonstration of particulate (gene) inheritance ran counter to the more general observation of blending inheritance; after all, most traits are governed by multiple sets of genes and appear to "blend" in the next generation. Not until geneticists began looking for definitive patterns of particulate inheritance was Mendel's work rediscovered and appreciated. Barbara McClintock's careful work in maize genetics indicated that some genes could "jump" from place to place within the genome. More than twenty years would pass before such transposable elements were found in other organisms and her discovery achieved wider recognition. These examples of ideas that could not gain acceptance at the time they were first proposed occurred in our own modern world of open scientific inquiry and unrestricted intellectual discourse. Imagine how difficult it was to begin to ask questions and do experiments in a society long accustomed to rigid traditions and inflexible religious authority. One might try to solve practical problems in these societies, such as building earthquake-resistant buildings in China, but questioning the veracity of traditional explanations was an altogether different proposition.

Alan Cromer argues that the Greek origin of Western science was a unique and highly contingent event which would have been impossible within strong Judaic, Christian, or Muslim societies—all dominated by a powerful clergy.[19] Within a soap-opera theology with a weak priesthood, and having a rich tradition of public debate, Greece made science a *public examination* of the real world, unconstrained by a religious or bureaucratic elite. This may explain why Muslim science faltered once the clerics came into power, and why Chinese science failed to advance. Moreover, "Greek geometry and speculative thought were unique inventions . . . ," claims Cromer. "Their analysis of planetary motion was not improved upon for more than fifteen hundred years."[20] By uniting abstract and mathematical reasoning with observation and analysis, the Greeks had initiated science in the West; but with the decline and fall of Rome, the Greek legacy was lost to western Europe.

Between the ninth and the fourteenth centuries the world's most advanced medicine, astronomy, mathematics, and natural philosophy were being practiced in the Arab-Muslim world, expanded from a Greek, Hebrew, and Hindu legacy. From the fifth to the fifteenth century, the world's most advanced technology was practiced in China. Yet science failed to move forward in either culture. By the fourteenth century, with the establishment of Muslim political hegemony, Koranic scholarship relegated the study of the natural sciences to peripheral irrelevance. China focused its attention on practical matters and made no effort to search for underlying laws or generalities. Despite carefully recorded observations over thousands of years, Chinese astronomers never developed a hypothetical system to explain planetary motions. They studied the stars so carefully to foretell the emperor's future. (We call that astrology, an activity that also has a long history in Europe.) Characteristic of authoritarian cultures in which knowledge is confined within the tightly proscribed territory of religious or bureaucratic elites, both Muslim and Chinese authorities forbade the use of Gutenberg's printing technology. Why was Europe so different?[21]

A major factor must have been the nature of western Europe's larger-scale geography itself. Divided by mountain chains such as the Pyrenees, Alps, and Balkans, Europe could not be tightly bound within a single empire as was China. At the same time, complex coastlines along both the Mediterranean and the Atlantic afforded safe harbors and convenient routes for vigorous maritime interchange. At a more local scale, ample rainfall sustained a host of useful rivers, while supporting agriculture and forestry throughout the region. These factors facilitated political fragmentation. Europe would develop the first major civilization *not requiring* a large, bureaucratic central government.[22] Scholars persecuted for their teaching in one state could march off to another. Ideas suppressed in Italy would invigorate the imagination of those elsewhere. A multitude of separate states and local autonomous principalities also promoted incessant warfare, which William McNeill sees as one of the "powerful wellsprings of the West's vitality."[23] Political division and conflict also contributed to less rigid societal stratification within Europe. In addition, a fundamental "structural" change had already taken place within European society, beginning in eleventh- and twelfth-century Italy.

## A LEGAL TRANSFORMATION IN EUROPEAN SOCIETY

Harold Berman claims that nothing less than a revolution, fundamental to the nature of modern western Europe, had taken place between 1050 and 1200. During that time, a succession of popes and a few small wars redefined the Roman Catholic Church as an independent and sovereign institution. No longer would bishops or priests be appointed by emperor, king, or prince. Usually referred to as the conflict over investiture, Berman calls this the "papal revolution."[24] He argues that this revolution inadvertently allowed a vigorous commercial society to divorce ecclesiastical law from mercantile and other legal traditions. It was at this critical moment in European history when a major repository of Roman law was rediscovered in 1080. The Roman legal tradition of more than a thousand years had been summarized and codified in 553 C.E. by a commission created by the Emperor Justinian. This compendium became the subject of intense study and active commentary by a succession of teachers at Bologna, and the school of Roman law at Bologna quickly became the center of a profound transformation. Because an expanding mercantile economy required more flexible legal procedures, Roman law was used to develop a practical and authoritative basis from which to derive those new procedures. Expanding commercial activity inspired well-to-do families to send their sons to study at Bologna—Europe's first graduate school of law. Not only did these studies create a new system of laws for the bustling trade of busy Europeans, these teachers of the law "gave the West its characteristic method of analysis and synthesis of texts," writes Berman. "They taught the West to synthesize cases into rules, rules into principles, principles into a system. Their method . . . was to determine what particulars have in common, to see the whole as the interaction of the parts. This was the prototype of modern western science. . . ."[25] Though we rarely think of the Western scientific tradition as being indebted to the Roman Empire, Berman's claim is insightful; legal practice and philosophy will inform all aspects of a society's cultural life. Rome's pragmatic genius, in the words of David Fromkin, had "manifested itself . . . in the development of a system of laws that remains one of the intellectual marvels of the human race."[26]

The papal revolution and the rediscovery of Justinian Roman civil law, within a dynamic mercantile environment, precipitated a subtle but powerful change. Will Durant wrote:

> In Islam, as in Judaism, law and religion were one, every crime was sin, every sin a crime; and jurisprudence was a branch of theology. . . . The papacy did what it could to halt the exhumation of a code that made religion a function and servant of the state, but the new [legal] study fed and expressed the bold rationalism and secularization of the 12th and 13th centuries. . . . St. Bernard complained that the courts of Europe rang with the laws of Justinian and no longer heard the laws of God.[27]

Separated from ecclesiastic law, innovative civil law opened European society to new possibilities. Toby Huff argues that the legal establishment of autonomous corporate bodies in "neutral zones" within European society were necessary prerequisites before private business and other enterprises could flourish.[28] Creation of the corporate entity (derived from the Roman *collegium*) treating collective participants as a single unit was a fundamental part of this legal revolution, allowing the development of a dynamic society free from overbearing religious or political control. Mercantile associations, craft guilds, and small "unions" of students or teachers were given special legal status as corporations. Being corporate entities, early universities were invested with important rights and protections. Because they often had no property, made a major contribution to the local economy, and could freely move elsewhere, universities managed the rare feat of securing local patronage and protection with only minimal interference, notes David Lindberg.[29] Thanks to this open social and legal environment, schools at Bologna, Paris, and Oxford became the first universities; by 1500, seventy universities had sprouted on European soil.

Universities provided the professional classes with a deeply shared experience in their formative years, helping the fragmented polities retain a singular cultural identity. Unlike earlier teaching institutions, and despite their association with religious or civil authority, the universities focused on a range of studies including philosophy, law, medicine, and natural sciences strongly influenced by Aristotelian precepts. Biblical literalism gave way to a search for first principles and primary causes. An unusual aspect of these universities was that they all shared the same texts and similar philosophical attitudes. In addition, they all communicated with a single language—Latin—allowing scholars to move freely throughout Europe. Better yet, as Jacques Barzun points out, "Logic as antidote to loose inference was helped in the Middle Ages by the use of the international language, not Latin, but *Medieval* Latin, a medium of exact expression, simplified syntax and enriched in vocabulary."[30] Marcia Colish claims that a university education was able "to guarantee a thorough knowledge of past and current interpretations of one's subject and the ability to analyze tradition with a keen and ruthless eye, giving no quarter to fallacies, irrelevancies, and outmoded views, however august the source."[31] Over time, the universities evolved from institutions that debated ancient wisdom to centers where new knowledge was actively pursued, helping western Europe sustain the open and public intellectual environment that would transform our world.

Though the rise of modern science was clearly a long and protracted process, deeply rooted in earlier Greek, Arabic, Hindu, Chinese, and Roman achievements, its full development demanded autonomous institutions within a free and open society. These institutions, the scholarly universities, were created *only* in medieval Europe and explain, in part, why no other culture in the world had been able to make a similar progression. Edward Grant sees this period not only as "one of the most glorious chapters in human history," but also as a fine example of multiculturalism.[32] First, Muslim scholars, deeply appreciative of earlier Greek, Hebrew, and Hindu science and mathematics, preserved and embellished that legacy. Then Europeans, eagerly absorbing the earlier traditions, further expanded these studies with great energy.

The infusion of Greek, Arabic, Hebrew, Hindu, and Chinese knowledge, an innovative philosophy appreciative of nature, a papal revolution allowing for the separation of the legal systems of church and state, a social arena open to independent institutions and busy trade, the discovery and exploration of distant shores, the printing press, and a profound religious reformation—all contributed to attitudes and ways of thinking that we now take for granted. Beginning in the medieval period and through the age of exploration, Europe created what Richard Tarnas has called "a new consciousness—expansive, rebellious, energetic, and creative, individualistic, ambitious and often unscrupulous, curious, self-confident, committed to this life and this world, open-eyed and skeptical. . . ."[33]

The growth and development of science in Europe was the outcome of a unique set of events. Amid an environment dominated by a powerful and intolerant Church (regularly burning dissidents at the stake), and despite incessant local warfare, important changes had taken place. Between the years 1000 and 1500, western Europe transformed itself from a technological backwater to the most advanced and dynamic culture the world had ever seen.[34] This new, relatively open society, powered by a well-organized trading system with strong legal protections, created a mercantile revolution that explored and exploited the entire planet, becoming the free-market capitalism that continues to change the modern world. Institutionally protected property rights and a developing respect for individual rights enhanced the environment for an intellectual revolution as well. After early medieval thinkers recognized nature as an acceptable subject for study, "the Western mind experienced an urgency, almost a demonic drive toward a more intimate understanding of the realm of matter through more careful observations and experiment, a drive to pass beyond the external barrier of the physical world, not into any spirit or psychic quality of things but into the inner material forces within things," declare Brian Swimme and Thomas Berry.[35] As so many times before in the trajectory of our long epic, this major cultural innovation was contingent on a complex combination of felicitous events; it was not simply a progressive efflorescence.

Whether here on planet Earth or elsewhere in the universe, the assumption that since science happened once, science ought to happen often is wishful thinking. The cultural and intellectual transformation that took place in western Europe was built upon knowledge and innovations first developed by Greeks, Chinese, Romans, Hebrews, Muslims, Indians, and others. These many different people had provided the foundation for the last major step in our odyssey. That advance has allowed us to unravel the deepest mysteries of nature and, in recent decades, to listen for signals from other stars. Today we enjoy the benefits of a modern industrial society at the same moment that we are unraveling the 4-billion-year odyssey that got us where we are. Finally, having surveyed an epic of more than four billion years, let's look at the more fundamental question: How rare might such a historical progression be? How likely is it that other technological civilizations are circling distant stars in our galaxy?

# 11

# PERFECT PLANET, BUT HOW UNUSUAL AN ODYSSEY?

clever species

Scientific inquiries over the last four hundred years have fashioned a new and extraordinary epic. Today nuclear physics contributes to the work of cosmologists as they develop theories that model an explosive birth for our universe. Geologists and planetary astronomers are creating a unified picture of how our planet and solar system have developed over the last 4,500 million years. Paleontologists, systematists, and molecular biologists are uncovering information that will help illuminate the historical panorama of life. Archaeologists, anthropologists, and historians are investigating the origin of humans and the diversification of our varied cultures over the last two million years in ever greater detail. Even as molecular biologists begin to understand how genes orchestrate the creation of a complex animal, they are also helping us understand the origins of life's diversity. Together, all these disciplines have contributed to our understanding of a vast and richly detailed epic, both cosmological and biological. A modern effort sharing some of the same aspirations as earlier periods, the unfolding scientific understanding of Earth's evolution and an expanded understanding of the cosmos, has been called "the major revelatory experience of our time."[1] That epic has informed these pages; it is our basis for attempting to answer an ancient query, "How unique might we be?"

Many scholars have insisted that our little corner of the universe has no special qualities, and, for this reason alone, we humans are nothing special. According to this "mediocrity principle," there ought to be a lot more like us out there in the galaxy. Haven't we learned that we are not at the center of the universe, nor at the center of our galaxy, and not even at the center of our own solar system? So it figures that since we are the product of the same forces operative elsewhere in the universe, we are, in effect, mediocre. Humbug! The purpose of these pages has been to confront the mediocrity proposition with a strong counterclaim: we humans are the product of uniquely felicitous events, and, though we are certainly unique in a specific sense (two legs, two arms, ten fingers, etc.), we may also be unique in a more profound sense, having developed the high technology and sciences allowing us to probe the mysteries of the firmament.

Consider the improbabilities. We start with a solitary star having a rocky planet orbiting at just the right distance to maintain water in its liquid state. This same planet has a nice big moon to stabilize its axial gyrations. Also, about three-quarters of its surface is covered by water—the result, in all probability, of having been pelted by a pack of comets. Add to all this a dynamic crust, lubricated by all that water, maintaining landmasses riding high above deep ocean basins. These unique attributes suggest that finding another planet as good as ours isn't just difficult; it's close to impossible.

Can we really take seriously speculation about sending forth colonies of humans to populate the galaxy? First, there is the fact that stars are located trillions of miles apart. Even if we could travel at the speed of light (671 million mph), it would take more than four years to get to the nearest star. And then there's Einstein's insight that material objects become more resistant to acceleration as they approach the speed of light, and those same objects are incapable of ever attaining or exceeding that speed. Furthermore, achieving speeds of 1 to 10 percent of the speed of light would require enormous amounts of energy. But that's only half the problem. Slowing down to stop and take a look around a distant planetary system requires that our spacecraft use *just as much fuel energy* as it did to reach the cruising speed it took to make the journey in the first place. Remember, brakes don't work in the near-vacuum of interstellar space. And getting back to Earth, or visiting a second star, will require just as much fuel all over again. That's a lot of gas.

Even if humankind were to discover some magical new cheap lightweight propulsive force, there's another huge problem: be sure that your spaceship doesn't run into *anything* unless it's going in the same direction at just about the same speed. Even cruising at speeds of "only" hundreds of thousands of miles per hour, running into something the size of a baseball might vaporize your spacecraft. (The energy produced by such high-speed collisions can be explosive.) Also, small, dustlike material will be peppering the front of the vehicle with powerful erosive force; be sure that your spacecraft has a continously renewable front end.

The fundamental problem of interstellar voyaging is the huge distances between stars. In 1978, Isaac Asimov generously estimated that there ought to be 530,000 technological civilizations in our galaxy, separated by an average of around 630 light-years.[2] Even if it were possible to travel at 10 percent of light's speed, 6,300 years adds up to a long ride. Obviously, space voyages to other nearby stars are not feasible until we learn how to freeze and unfreeze ourselves or build intelligent robotic surrogates. Once again, consider these major factors: tremendous fuel requirements for high-speed acceleration and deceleration, huge distances between stars, and the difficulty of sustaining human life *and health* on gravity-free flights covering hundreds of years. Then there's the problem of surviving on distant Earthlike planets lacking water and free oxygen. But let's say you're lucky enough to reach a planet with liquid water, land surfaces, *and* oxygen. The purple slop that helped put those oxygen molecules into that distant planet's atmosphere might find the moist interior of your lungs a great place to party. What's the likelihood of your immune system, developed over 300 million years on our own solitary sphere, being able to deal effectively with a life-form it had never encountered before?[3] Just a few of these considerations would make one think that discussions of interstellar travel would be restricted to science fiction. Not so.

Although science fiction writers and a few enthusiastic astronomers claim that our galaxy probably has many civilizations more advanced than ours, a number of academics have argued that *if such super-civiliza-*

*tions existed*, they would have populated all the appropriate planets in the galaxy by now. After all, the universe is probably somewhere between 12 and 15 billion years old, plenty of time to evolve intelligence and high technology. Some of these pundits think it's just a matter of time before we ourselves learn to build a "self-replicating universal constructor with intelligence comparable to the human level." This glorious product of the human imagination "is a machine which is capable of making any device, given the construction materials and a construction program." Provided with human intelligence, this fully outfitted contraption would be able to sally forth on interstellar rocketry, to search out construction material on the planets of a distant star; there it would make copies of itself and its vehicle. These machines and their newly minted progeny would then travel to further stars, where they would continue the process. How such amazing "von Neumann machines" would acquire the energy to keep themselves going is a question that hasn't been answered. Also, there's not much thought spent on the problem of mining distant planets that have no tectonically concentrated ores, or how you build new spaceships in acid-laced atmospheres lacking oxygen.

Adrian Berry, in a book that mixes fact and fantasy, proclaims that "even a small asteroid five kilometers [three miles] across contains the wherewithal to build several hundred space ships."[4] (Haven't these guys ever seen a blast furnace in action—or visited a manufacturing facility? Building a spaceship takes a civilization.) Since these authors believe that space probes are easy to assemble, they have argued that there probably aren't any other super-intelligent civilizations out there—because they haven't visited us. This is just one example of the kind of creativity that has fueled speculation about other high-tech civilizations in our galaxy. Another idea is that there are *bad* technological civilizations in the galaxy that have been busy destroying other colonizing civilizations—and that's why we haven't been visited yet. Please note that these views have been published by scientists in scientific journals—not by the writers of science fiction in pulp periodicals.[5]

Surely, the near-impossibility of interstellar travel by heavy, mechanically vulnerable, space vehicles makes the lack of space visitors a simple consequence of the nature of our universe. "The workings of biology, the physical laws of energy and the vast interstellar distances make interstellar colonization unthinkable for all time," declared radio astronomer and respected SETI pioneer Frank Drake. His position is, I believe, bluntly realistic.[6] So let's forget space visitation and take a more reasonable approach to the question of how many other advanced civilizations might be in our galaxy.[7]

## HOW LIKELY ARE OTHER TECHNOLOGICAL CIVILIZATIONS IN OUR GALAXY?

To quote Frank Drake again: "The chance of any earthly process being unique in the universe is surely the most unlikely of all possibilities," and I'll wager most everybody can agree with this sentiment.[8] It is an expression of the "cosmological principle" declaring all places in the universe to be fundamentally alike in regard to the laws of physics and chemistry. This is the philosophical foundation from which we attempt to understand the distant universe, and it seems to work just fine. But the problem of our uniqueness is not one of process; it is one of coincident and necessary historical circumstances having allowed various processes to reach their full development. In discussing the probability of other technological societies, Drake developed what has come to be called the Drake Equation.[9] The equation is nothing more than a series of factors that, taken together, were used to discuss and estimate the probability of far-distant civilizations. This was part of early efforts to organize a search for radio-frequency evidence of extraterrestrial intelligence. None of the seven probabilities or factors can be determined with any degree of precision, but they form a convenient outline for discussion.

Having surveyed how one species managed to get into the business of building radio telescopes on our planet, let's explore the probabilities of other planets with radio telescopes by using the Drake Equation. This equation estimates the number (N) of detectable civilizations in our galaxy by multiplying the rate of star formation (R) times the fraction ($f_p$) of stars that form planets times the number ($n_e$) of planets hospitable to life times the fraction ($f_l$) of those planets where life actually emerges times the fraction ($f_i$) of planets where life evolves into intelligent beings times the fraction ($f_c$) of planets with intelligent creatures capable of interstellar communication times the length of time (L) during which such civilizations remain detectable.

### 1. The rate of star formation in our galaxy

The first of Drake's factors is an estimate of the rate of star formation (R). Ken Croswell estimates that the current rate of star formation in our galaxy is about ten stars per year,[10] while Andrew and David Clark estimate the number at twenty.[11] However, it is very likely that this rate has decreased over the last 10,000 million years. Let's turn things around and replace this parameter by starting with the current population of stars in our galaxy, which is estimated to be as high as 300,000 million. We'll then use the other factors in Drake's equation to chisel down this figure.

Three hundred thousand million stars is a lot of stars, but this number must be constrained by the requirement for a very stable long-lived star. Large fast-burning stars don't last long enough to allow life to achieve higher levels of cognition. And, as we discussed earlier, the problem with small stars is twofold. First, they are often unstable, flaring up from time to time. Second, a small star will have a very narrowly constrained Goldilocks orbit, with low probability of having a rocky planet at precisely the right distance. And because such an orbit would have to be close to its cooler star, there is a strong probability of having the planet's rotation slowed or locked by tidal forces. When we restrict our search to Sunlike stars, we run into the problem of doublet and triplet star systems. In these cases, a stable circular orbit for an Earthlike planet is highly unlikely. Thus, with only about 10 percent of stars belonging to the appropriate size classes, and only half of these stars solitary, we've already knocked the estimated figure of 300,000 million down to the neighborhood of 15,000 million.[12]

### 2. The fraction of stars that form planets

The second term in Drake's equation is an estimation of the percentage of stars that form planets ($F_p$). This rate may be very high for solitary Sunlike stars, probably a factor of 1 (100 percent). However, our estimated 15,000 million stars may include many stars having formed in regions of low metallicity, where it is unlikely that inner rocky planets of substantial size will be formed. The requirement of having lots of "star dust" for our inner rocky planet will lower this probability. Also, there's the problem of having a trajectory, over thousands of millions of years, restricted to quiet regions of the galaxy, where being blasted by high radiation and supernova explosions is unlikely. (Recall the "habitation doughnut" we discussed in chapter 1.) Putting the metallicity and safe-neighborhood requirements together, we'll now guess one out of five; so we're down to 3,000 million stars.

### *3. The number of planets hospitable to life*

With this third factor, we run into the troublesome "Goldilocks" problem. Contrary to Drake's optimism, it is highly unlikely that any star will have more than one planet at "just the right distance." However, some astronomers have recently expanded the potential sphere of life out to beyond the orbit of Jupiter. They do this because tidal forces within the moons of the giant planets might generate enough heat to keep water fluid under an icy surface; perhaps that's what's going on under the surface of Jupiter's moon, Europa. Such an environment might support bacterial slop in a dark liquid interior, but it won't give rise to complex life-forms that are hungry for energy. (Deep under ice, there's not a lot of energy available.) So let's stay with the more classical concept of the Goldilocks orbit.

Unfortunately, energy is only the beginning of the Goldilocks problem. In addition to being at just the right distance from its star, this lucky Goldilocks planet mustn't be too massive or too small. If too massive, quick-witted terrestrial life cannot evolve; if too small, a protective atmosphere will soon evaporate into the void. Our suitable planet has got to have a lot of water for life to prosper. Remember, too, the planet shouldn't have a suffocating, poisonous, Venuslike atmosphere, or too eccentric an orbit, or too slow a spin. Then there is the newly recognized possibility of having "wayward Jupiters" drift inward, closer to their stars, eliminating the possibility of an Earthlike planet in the Goldilocks orbit altogether. In fact, the discovery of so many different new solar systems in nearby stars strongly indicates that planet formation is often very chaotic. However, this suggests the possibility that a Jupiter-sized planet in the Goldilocks region might possess Europa-like watery moons on which life might flourish. Though capable of sustaining aquatic life forms, such smaller bodies are not likely to sustain an atmosphere dense enough for life on land.

Considered together, a stable system such as ours, with a wet planet of appropriate size in just the right orbit, may be rare indeed. And since we are looking for complex life-forms, not just bacterial slop, Europa-like environments will not suffice. All told, one in one thousand star systems might be grandly optimistic from our point of view, and now we're down to the probability of 3 million stars in our galaxy.[13]

### *4. The fraction of planets where life actually emerges*

If we consider how quickly simple life-forms developed on our planet, it seems likely that this factor has a very high probability. Drake estimated this factor to be 1 (100 percent), and I would agree. Of course the real kicker here is tied to the previous factor: *suitable* planets with all those Goldilocks parameters. Unless all the "just right" conditions are present, this factor has zero probability, at least in regard to larger complex life-forms.

In our solar system, planet Earth is unusual in that it experiences *two* forms of continuous energy flux that can sustain life. The most important, of course, is the energy we receive from the Sun. Because we're in the Goldilocks orbit, the Earth's surface is showered with energy in just the right amount and intensity to power life. However, it is the second source, at deep sea volcanic vents, where concentrated high energy flows outward into cooler aqueous surroundings, which may have been critical for the actual origin of life. Here a richly varied chemical soup could boil and bubble over millions of years within a larger environment of cold deep-sea waters. Perhaps this was the critical factor for the relatively quick appearance of life on planet Earth. So let's keep things simple and assume that all the "suitable" planets of the previous factor also had such conditions. This leaves the probability of life's initiation at 1 (100 percent), and we're still dealing with an estimated 3 million stars with life-supporting planets in our galaxy.

### *5. The fraction of planets where intelligent beings evolve*

The problem with this factor is that different levels of intelligence have very different probabilities of developing. In the case of planet Earth, it took 4,000 million years before complex animals arose, and virtually all of them possess a modicum of what we might call intelligence. Really high intelligence is another matter entirely. Of the tens of millions of species that have inhabited Earth, living and extinct, only one terrestrial species has developed a really powerful brain. Living in the sea failed to promote the escalation of intelligence; it seems likely that exceptional intelligence requires the challenges of a richly complex and challenging land environment. Whales and dolphins are exceptional; they are mammals that have *returned* to the sea. A terrestrial environment is essential for another obvious reason; despite their big brains, dolphins and whales aren't about to develop metallurgy anytime soon.

Finding a comfortable *terrestrial environment* on a distant planet may be highly improbable. First, there is the need for lots of water to keep the winds moist and rains moderately reliable. Then there is the problem of vegetated land area of considerable breadth on which an intelligent lineage can evolve. On Earth, the two-tiered surface of deep ocean basins and elevated continents has been created and maintained by plate tectonics. On a distant planet lacking such tectonic activity one might imagine land surfaces slowly eroding into the sea, and their land biota surviving only by leaping from old eroding volcanoes to those newly forming. Planets with lots of water *and* land surfaces that are stable over hundreds of millions of years may be extremely rare in this and any other galaxy.[14]

Many people seem to think that, once a rich land biota is in place, really big brains are just a matter of time. Drake's belief that dinosaurs with our general height and weight were "well on their way to developing intelligence" is, in my opinion, way off base.[15] In a reign of over 150 million years, only one late group of predatory dinosaurs (*Troodon* et al.) developed larger brains. Speaking of dinosaurs, I'm reminded of that long historic interlude during which the mammals weren't going anywhere. The ruling reptiles, including the most awesome carnivores in Earth's history, dominated the land for over 100 million years. If they hadn't gotten zapped, we, in all likelihood, would not be here. The mammals were unable to diversify and proliferate while dinosaurs ruled the Earth. How many other planets might be ruled by similarly ferocious but dim-witted predators?

Armand Delsemme contends, "Since the brain changed so fast, its recent evolution cannot be a major bottleneck in the evolution of intelligence."[16] Earlier, I argued that an organ as "expensive" as our brain could be built only under intense and continuous selection. In our case, that selection was provided by a unique within-species arms race. If we had not evolved, only our closest cousin, the common chimpanzee, seems to have the behavioral traits that might someday lead to a similar trajectory. At this time, no other lineage of land animals, out of thousands, seems capable of duplicating our achievement. (If elephants had fingers, things might be different.) Contrary to what many seem to believe, natural selection does not exhibit an inherent progressive drive. Complexity took a long time to evolve. Really big brains are rare; they were the outcome of very special circumstances. Our big brains didn't just happen; our genome had to be shoved forward with continuing and relentless selection. Building extraordinary brains required extraordinary pressure.

Amir Aczel, writing about the question of distant civilizations from a statistical point of view in *Probability 1*, ventured the opinion that "we probably are among the most advanced civilizations in our Galaxy."[17] At the other end of the spectrum, Carl Sagan, in *Pale Blue Dot*, imagined that the universe is filled with beings far more intelligent than we.[18] But why should that be? Consider first that there is no evidence for human beings having become more intelligent over the last 20,000 years, or that we are getting any smarter today. Without continuing strong selection, there is absolutely no reason for a continuing trend toward greater intelligence. In modern Western societies, large numbers of smart and creative people are among those with the lowest birth rates; this is negative selection. Some commentators have suggested that the increase in cesarean-section births will allow us to evolve bigger brains. Not so. A society that tries to save every child, however severely compromised, is not likely to develop better brains. Considering the extraordinary accomplishments our present onboard computers have been able to accrue, it seems highly questionable that there should be even higher forms of intelligence in the distant universe. Unless, of course, those faraway superintelligent beings have engaged in a focused program of selective breeding or genetic engineering.

Let's not forget that there's an astronomical factor that's helped set the stage for a big brain on planet Earth: a fat moon. Without such a stabilizing satellite, a badly wobbling planet will put huge stresses on terrestrial vegetation and the animals it supports. It seems unlikely that we would've made it without our beautiful satellite. Wobble-free rotation, lots of water, large stable landmasses, and complex land vegetation are not likely to be common in this or any other galaxy. But that's not all. There's another astronomical factor. It's important to have large, stable outer planets in near-circular orbits to sweep up or eject potentially deadly planetary debris that might otherwise keep colliding with our Goldilocks planet. One conclusion seems inescapable: even on planets that have all those qualities that can support simple life-forms, the evolution of a level of intelligence that can imagine the future and fashion elaborate myths to explain the mysteries of nature has a very low probability. Here on Earth, this superior level of intelligence required a planetary mass that allowed quick and clever creatures to evolve, a near-circular orbit, days that are not too long or too short, and large stable land surfaces supporting moist, equable climates over hundreds of millions of years. As with the previous factors, there is no way to figure these probabilities precisely, so I'll guess one in a thousand. Starting with the 3 million stars orbited by lively planets, that gets us down to only about three thousand with really intelligent beings.

### *6. The fraction of planets with creatures capable of interstellar communication*

Finally, we've reached the stage of building radio telescopes by a fully developed scientific-technological, energy-rich society willing to spend big bucks on SETI-like activities. This factor demands a lot more than smart hunter-gatherers or even larger communities with shamans and politicians. Again, astronomers and many others seem to think that once a really intelligent species has evolved, building radio telescopes is just a matter of time. As in the case with the development of high intelligence, these views reflect a deep belief in an underlying progressive direction to cultural history. This attitude is contradicted by the fact that only one culture, out of thousands, has created a society supporting a major revolution in how humans analyze the world around them. A fine onboard computer was not enough. The "cultural software" of the scientific method was needed to develop experiments and instruments that would extend our vision far beyond our senses. The sci-tech revolution was not the result of an unfolding manifest destiny; it seems instead to have been the product of yet another series of glorious accidents in our long odyssey. Those accidents included agriculture, metallurgy, and the industrial revolution. Given enough time, how likely is it for really intelligent cultures circling distant stars to break out of their religious, military, or bureaucratic conservatism and focus on deep analysis of the world around them? Building an efficient bureaucracy is very different from developing a tradition of careful experimentation and the testing of hypotheses.

A more fundamental aspect of scoring this probability brings us back to the Goldilocks planet again. One of our arguments has been that without plate-tectonics, providing elevated continents and metal-enriched ores, there would be neither large persistent land surfaces *nor metallurgy*. But we may have included this requirement in our previous factors when estimating the probability of stable continents on planets with lots of water. Those earlier factors, however, did not include another terrific lucky break: the flowering plants. This unique and dynamically progressive assemblage promoted the evolution of smart climbing monkeys, filled the savanna woodlands with fruits, nuts, and tubers, and, finally, provided us with rice and beans. Even today, after 10,000 years of agriculture, about twenty-five flowering plant species provide us with 90 percent of our vegetarian food energy—that's twenty-five out of 250,000 species. Without this major photosynthetic lineage, so rich in species and diverse in useful tissues, there would have been little probability of developing higher civilization or radio telescopes on planet Earth. Concluding his essay "How Flowers Changed the World," Loren Eiseley phrased it more elegantly: "The weight of a petal has changed the face of the world and made it ours."[19]

If we review these various parameters together, and being optimistic once again, we may guess that the probability of developing this last telescope-building stage ranges somewhere between one in one hundred and one in one thousand; and now we're down to the neighborhood of three to thirty Earth-like planets with technological civilizations in our galaxy.

### *7. The length of time that civilizations remain detectable*

Some scientists who have wrestled with the Drake Equation have concluded that the huge number of stars in our galaxy and the great improbability of having all those necessary conditions for really intelligent life to evolve balance out (N = L), so they place special significance on this last factor in the Drake Equation. Some of these authors have concluded that if the lifetimes of communicating civilizations range between one thousand and one million years, *then the numbers* of other civilizations out there in the galaxy should also range between one thousand and one million. It would seem that none of these experts has looked at the unusual aspects of our planet and our history in the same way we have. Nor do they seem to be fully cognizant of what's going on here on Earth right now, all around us.

The last of Drake's factors, the lifetime of communicative civilizations, must surely be the most disheartening for SETI enthusiasts. Thanks to the single most significant intellectual revolution in the history of mankind, our accomplishments now threaten the planet's biosphere itself. Second only to the invention of agriculture in effecting social change, the scientific-technological revolution has torn humankind loose from its ecological moorings. Drake's last factor requires a deeper look; a not wise but clever species is transforming the surface of our planet.

## A SPECIES OUT OF CONTROL

Humans, like other species, probably have been held in check in their homeland by ever-present diseases, parasites, and occasional famine. Then, moving to new landscapes with fewer pathogens, we expanded across the planet in a way no other large mammal had before. Agriculture transformed that expansion into an explosion. With science, medicine, and technology (and all our age-old habits unimpeded), we are currently in the business of turning that explosion into a planet-sized disaster. Humanity is riding the crest of the sixth major extinction event in the history of complex life on planet Earth—and we are its fundamental cause.[20] India now numbers a billion people, having *tripled* its population in only fifty years; an achievement that has been characterized as "the most successful program of poverty multiplication in human history."[21] Our planet is witnessing a continuing growth in our overall population by around seven thousand *more people every hour*. This statistic does not bode well for the future, but few people show any concern. In fact, journalists and pundits dismiss the problem by trumpeting figures showing the rate of population growth to be *decreasing* all over the world.

Declining birth rates are real and widespread. The worldwide annual growth rate peaked at 2.1 percent between 1965 and 1970 and has been declining ever since; annual growth reached its maximum in 1996, with somewhere between 92 and 95 million additional bodies.[22] Unfortunately, continuing growth is substantial. This may seem like a contradiction but it is not, thanks to two important demographic factors. The first is simple: there are *twice* as many people on our planet today as there were in 1965 to 1970, allowing a lower growth rate to produce almost as many additional people. The second factor is a bit more subtle: a significant percentage of the world's young women are just now entering their reproductive years, and, as they begin to give birth, they counterbalance a declining birth rate; this is called "demographic momentum." The consequences are simple: there is only a slight decline in the number of people actually being *added* to the planet each year. The AIDS pandemic was forecasted to claim as many as 3 million lives during the first year of the new century, creating millions of orphans and a devastating economic impact as it spreads. As horrible as this disease is, the numbers lost have only a small effect on the overall growth rate of the world's population. The human population will continue to grow by more than 70 million people a year for some time to come.

By improving living conditions, expanding the availability of food, and understanding the nature of disease, we have broken loose from the factors that regulated our numbers over previous millennia. "Scourges, pestilence, famine, earthquakes, and wars are to be regarded as blessings, since they serve to prune away the luxuriant growth of the human race . . . ," claimed Tertullian in second-century Carthage. Few people care to experience such "blessings," and we have striven mightily to thwart those factors that once contained us. The scientific, medical, technological, and "green" revolutions have combined to make our modern times especially historic. I would argue that two world wars, the threat of nuclear holocaust, and the information explosion were peripheral to the central historical dynamic of the twentieth century. The central dynamic of the last one hundred years is that the human population ballooned from 2 to 6 billion and continues to inflate.

Most businesspeople, some economists, most politicians, and virtually all religious leaders seem to think of population growth as a truly glorious development: more customers, more taxpayers, more voters, more souls. Most journalists and writers discuss the future of our society with scant reference to our increasing numbers. In contrast, biologists and ecologists view the growing numbers with profound dismay. From a biological point of view, it looks as if *Homo sapiens* will do what all other organisms on the planet in a similar position have done. We'll continue to grow exuberantly, utilize every resource we can get our hands on, and have a really great time. Then, as resources become scarce and the biosphere deteriorates, our technological juggernaut will falter. Here's a common pattern in nature, where populations overshoot the carrying capacity of their environments when times are good and then suffer a population crash. This is a cycle of boom and bust. Though most such systems recover, they end up at levels well below the "good times." Can our industrial-information society, built on the presumption of continuous growth, avoid a "bust?" Not likely.

In a superbly referenced survey of our twentieth-century achievements, J. R. McNeill points out that "many of the ecological buffers—open land, unused water, unpolluted spaces—that helped societies weather difficult times in the past are now gone."[23] And when the bust does come, it's unlikely that our planet will be able to sustain us in the manner to which we've become accustomed. But not to worry. As biodiversity declines and ecosystems contract, one larger mammalian species will survive, together with roaches, ragweed, and rats. *Homo sapiens* is simply too damn tough. If we can survive and reproduce over generations at the edge of Arctic ice floes, in the desert thorn scrub of the Sahel, or on the back streets of Calcutta, we'll muddle through.

Most ordinary citizens, no matter where they live, as well as journalists, economists, religious leaders, and politicians, seem utterly unable to comprehend the full implications of exploding human populations. It's just not a problem. "Perhaps because it is unprecedented, the environmental crisis seems completely beyond our understanding and outside of what we call common sense," writes Al Gore in *Earth in the Balance*.[24] Some of us had thought that famine would accompany the continued expansion of human numbers. Except in regions of political turmoil or unusually bad weather, famines have indeed been averted. Both the expansion of time-honored traditional farming and new fertilizer-augmented agricultural practices have met the challenge of feeding exploding populations. We had also dreaded a massive nuclear exchange between warring superpowers. Today, we are faced with something more insidious because it is multifaceted and difficult to see: deterioration of agricultural soils, falling water tables, declining ocean fisheries, widespread dispersal of invasive species and pathogens, deforestation and overgrazing, rampant urban expansion, increasing unemployment, escalating immigration—all aggravated by a relentless increase in human numbers.

The late economist Julian Simon and his associates viewed the world very differently. They claimed that virtually everything is getting better, and the data they brought together make a very strong case for such an optimistic view.[25] No matter whether you measure the last few decades or the last four hundred years, health *is* better, violence *has* decreased, personal property *is* greater, and democratic governments *are* more common. Trouble is, Dr. Simon and many other economists seem to have an extraordinary faith that some new discovery or new invention will solve any and every problem that might arise. But the real world offers some sobering statistics. Because of venture capitalism, the world's fishing fleet grew by 90 percent between 1970 and 1992. With improved technology, vessels can now catch more fish faster, and all but one of the ten major commercial fish populations of the world are in decline today. Will people develop a sustainable harvesting program for the world's oceans? The problem is formidable: people need to catch fish, make money, repay their debts, and feed their children today—not tomorrow.

According to Simon and many other economists, alternatives will be found, new solutions will be developed, and things will only be getting better and better. A similarly optimistic review published by the U.S. National Academy of Engineering claims that we will be able meet the challenges of a planet that may have as many as fifteen billion people.[26] But these are all hopeful scenarios for a self-propelled phenomenon that is not the product of consensual global planning—it's just happening.

What is especially troubling is that explosive population growth is not even seen as a problem worth discussing, and most people are completely unaware of its effects. Julian Simon and his associates failed to mention that there are *many millions more* people undernourished today than there were one hundred years ago. India's gallop toward a billion citizens merited only an occasional aside in the newspapers. Conflicts in the Balkans or in Burundi are always analyzed politically and culturally, *never demographically*; birth rates in Kosovo or Afghanistan are not part of journalism's field of vision.

Recently, the editors of the Sunday *New York Times Magazine* published an article by Ben Wattenberg titled "The Population Explosion Is Over."[27] Wattenberg, in the manner of Simon, focuses on real and reasonably accurate data regarding a number of countries in which population growth is below replacement numbers. Although his facts are in good enough order, there is a veritable mountain of data he ignores. Wattenberg fails to mention such unpleasant facts as Egypt's population growing by a million people every ten months and Mexico's by 2 million every year, or that the number of people in Nigeria, Africa's most populous country, will double in twenty-five years. While focusing on those few countries with unusually low birth rates, he fails to mention that in many third-world countries *half* of the population has not yet reached reproductive maturity. China has taken draconian measures to curtail its birth rate and yet *continues to increase* its numbers, due to the growing numbers of young women now entering their childbearing years. Over the entire planet, ongoing demographic momentum will keep human population growing rapidly for some time, despite declining birth rates. Wattenberg, a senior fellow of the American Enterprise Institute, makes no mention of demographic momentum. He is clearly singing the song of unfettered capitalistic growth: there can't be too many customers.

Even more troubling is the chorus of criticism regarding China's severe efforts to control its expanding population. Those waving the banner of individual rights and insisting that no religious freedom be abridged are the most outspoken in their attacks on the government of China. Zealous advocates of human rights seem to have no comprehension of the enormous problems facing the world's most populous nation. The possibility that our most cherished values of personal and religious freedom might someday propel us over the Malthusian edge like an army of lemmings headed for the sea doesn't seem to be worthy of discussion.

Utilitarian philosopher Lincoln Allison, in his otherwise sensible book, *Ecology and Utility*, labels objections to the concept of unlimited growth as "Malthusian error; there is no reason to suppose that any growth rate cannot be sustained. . . . We are always running out of resources and substituting others for them."[28] Likewise, Peter Huber declares that "The only limits to how much food we can grow, energy we can extract, houses we can build, miles we can travel, pigs we can breed, diseases we can cure, are the limits of human ingenuity."[29] Unfortunately, a huge majority of biological scientists cannot take such pie-in-the-sky optimism seriously. We're more in line with historian Paul Kennedy who writes, "It is inconceivable that the earth can sustain a population of ten billion people devouring resources at the rate enjoyed by rich societies today."[30] And, as if our rampantly increasing numbers weren't bad enough, our richest societies are engaged in an escalation of conspicuous consumption that Robert Frank calls "luxury fever" (sport-utility Cadillacs come to mind).[31] Ready to follow in these footsteps are billions of the world's "disadvantaged," eager to help pump additional $CO_2$ into the atmosphere. Concluding their historical survey of science and technology, James McClellan and Harold Dorn state, "The Industrial Revolution and its consequences have transformed historical circumstances so rapidly and in such profound ways that the current modes of intensified industrial existence are not likely to continue."[32]

With population growth declining only slightly, an insatiable demand for energy-rich lifestyles for everybody, and an inability to forge long-term, strictly enforced social strictures on a society whose principal ethical value has become unrestricted individualism, the currently dominant technological civilization on planet Earth may not have a long future ahead of it. Sure, our species will muddle on, but for how long will we be able to sustain our extravagant lifestyles? And why should this be a uniquely human dilemma? Once freed of constraints, wouldn't high-tech civilizations in other corners of the galaxy also be likely to experience uncontrolled growth, foul their environments, and make a mess of things? Surely, we can invoke the cosmological principle again: if Murphy's law applies here on Earth, it applies everywhere.

Garrett Hardin, whose 1968 essay "The Tragedy of the Commons" helped focus attention on the deterioration of our shared biosphere, thinks that the only way we humans can avoid ecological disaster caused by uncontrolled population growth is through "mutual coercion, mutually agreed upon."[33] Considering the religious, political, and individual priorities that must be dealt with, it may be impossible to reach a self-regulated, sustainable relationship with our earthly environment *before* major calamities bring about a forced contraction of our numbers and our extravagant lifestyles. Laurie Garrett, in a detailed survey of recent disease epidemics, fears that a crowded planet with rapid intercontinental travel is uniquely vulnerable to a catastrophic plague.[34] But a catastrophic reduction of human numbers might, sadly and paradoxically, be in the best interests of the Earth's biosphere—at least in the short term.

Getting back to the "L" in the Drake Equation, the present drama unfolding on planet Earth makes it seem highly likely that energy-guzzling technological societies have only a short life span; and perhaps that's just one more reason why we haven't been hearing from anybody on the interstellar airwaves. Thousands of hours and millions of dollars of listening time on dozens of giant radio telescopes have not revealed any coherent signals that might reveal the voice of a distant civilization. A short life span for technological civilizations and the many improbabilities of the Drake Equation clearly imply that if there are other technological civilizations in the galaxy, they are few and far between.

## MIGHT THE HUMAN ACHIEVEMENT BE UNIQUE IN OUR GALAXY?

The paper of your lottery ticket may be ordinary, the printing by automated machinery, the purchase no different from tens of thousands of others, but if it has the numbers that match those of the million-dollar jackpot drawing, you have something very precious. We and our dear planet have come a long way. Gravity and all the other general rules of physics, chemistry, and biology surely obey all the same rules here that they obey everywhere else. The processes that have played themselves out over these last 4,000 million years may have required no miraculous assistance. But the fact is that we've ended up in a very special place and time. This wasn't just a winning ticket; we've had a whole string of winning tickets.

Is life in the universe common? It would seem that while the probability of wet, rocky planets at the right distance from their stars is not great, there must be many among the hundreds of millions of Sunlike stars in thousands of millions of galaxies. But the majority of such life might not be much more than bacterial slime in moist depressions,[35] or jellyfish quietly drifting through their watery worlds, with dim-witted creepy-crawlies a rare exception. Recall that 4,000 million years had passed before our perfect planet became the home to complex animals. Yet, once the ball got rolling, the parade of life on Earth was much more than a random walk from simple beginnings. As we saw earlier, modularity, complexity, and increasing intelligence were trajectories driven by selection, chance, and necessity.[36] Nevertheless, the difficulties of developing a rich terrestrial biota, a really superior brain in at least one species, agriculture, metallurgy, and the social environment to encourage the serious study of nature, render the probability of other technological civilizations minuscule indeed.[37]

One would think that all the wonderfully improbable contingencies that have peppered our past might be seized upon by the religiously inclined as clear evidence for the unfolding of a providential plan. Glued down to a six-thousand- to ten-thousand-year-old universe, so-called creation scientists are unable to incorporate many recent scientific revelations into their thinking. However, for those who are not fixated on biblical "inerrancy," the nonstop trajectory of lucky breaks that seem to have played themselves out over 4,500 million years on planet Earth may indicate a divine agenda, especially the quick appearance of life after the late heavy bombardment, and that business of tripling hominid brain volume inside three million years. And here we are today, busily heating up the world, burning fossil fuels, just in time to fend off the next glacial cooling cycle.[38]

Atheists will counter that without such felicitous events we simply wouldn't be here, and, among the 300,000 million stars of our galaxy, somebody was bound to have gotten hold of a string of winning tickets. Regardless, these are all untestable metaphysical speculations that lie outside the purview of science, and we seem to be digressing.

However unlikely our odyssey, the incontrovertible fact is that our planet, our solar system, and our star are ideally configured for the development of intelligent life here on Earth. And it's not only our star and its entourage having been "just right" for our sustenance; it turns out that the entire universe is also "just right." Fred Hoyle pointed out that the stable resonance level of carbon 12 is a uniquely significant attribute of our universe; if it differed even slightly, life based on carbon could not exist. That quality of carbon depends upon one of the basic nuclear forces having its exact value and no other. Likewise, if gravity or the speed of light had slightly different values, our universe would be very different, and creatures like us wouldn't be here to wonder about it. Nor could we exist in a universe of different dimensionalities. This concatenation of felicitous attributes is the basis of the "anthropic cosmological principle." In its weak form, the principle is simply the recognition that many of the observed properties of the universe must have the precise values that they do because if they had other properties we simply wouldn't be here to study them.[39] Using the word *anthropic* is, of course, anthropocentric; in a universe with different properties, there wouldn't be any bears or butterflies either.

Along the lines of the anthropic principle, some scientists like Paul Davies find that "The physical universe is put together with an ingenuity so astonishing that I cannot accept it merely as brute fact. There must be, it seems to me, a deeper level of explanation."[40] That "deeper level" or suggestion of purposeful design may be impermeable to scientific inquiry and more a matter of belief. Here is where religious philosophers and the scientifically inclined find themselves in close quarters. Among these there are scholars who promote the "strong form" of the anthropic cosmological principle, claiming that the universe has its very special qualities *in order that* we and other sentient beings would come into existence at some stage in its history. Michael Denton's *Nature's Destiny* is a well-referenced argument for this point of view.[41] But such a bold claim, for a universe specially designed to put us here, is not only clearly untestable, it is also indistinguishable from a religious belief. That said, we've got to admit that if we really do live in an anthropic universe, as so many religious traditions claim, then there may be a lot more like us out there—and maybe some have even been visiting. Alternatively, if God simply kick-started the universe and let natural law take it from there (the sort of stuff science can investigate and analyze), then we're following the scientific agenda pretty closely and we can leave purposeful scenarios for others to explore. So let's get back to our subject and, once more, consider the question of how common technological societies might be—from within the constraints of a strictly scientific perspective.

In a universe lacking purposeful providential supervision, there are, perhaps, six fundamental hurdles in the development of technological civilizations. The first is the Goldilocks problem: having a warm, moist, quickly spinning Earth-size planet in a circular and stable orbit around a long-lived, stable, solitary star. That planet must develop a metabolically friendly atmosphere and maintain those conditions for thousands of millions of years. The second hurdle is developing a form of life that can not only replicate itself but also develop at least one lineage with the essential ability of transforming the energy of warm starlight into "food" available to the entire ecosystem. Once photosynthesis is in place, evolution can ignore the second law of thermodynamics and start crawling uphill.

Third, at least one lineage must have flexible thin-walled cells which can function together to develop large multicellular, energetic, and sentient animal-like forms. Fourth, a photosynthetic lineage of plantlike organisms must adapt to life on terrestrial surfaces and form a species-rich, three-dimensional vegetation. Fifth, a terrestrial animal lineage must develop a body plan and lifestyle in which at least one species drives itself into the business of building an expensive brain.

Finally, that lucky superintelligent species must have a generous physical environment in which it can develop agriculture, metallurgy, and energy-hungry industry, as well as a social environment supporting the continued pursuit of scientific inquiry. It goes without saying that this superspecies must have had keen vision, flexible limbs, and dexterous digits to fashion everything from primitive tools to telescopes. All this on a planet that has been circling its uneventfully reliable star for thousands of millions of years. And so, after having briefly considered the Drake Equation and having reviewed our own long odyssey, we come back to the question: how many planets are the home to intelligent beings who have transcended all these hurdles to produce sophisticated scientific societies in our own galaxy?

Perhaps, at this moment in time, there's only one.

If we consider all the lucky breaks in our extended epic, it's not too outrageous to claim that maybe we've got the only radio telescopes in the galaxy. But this is just the Milky Way galaxy; there may be as many as 500,000 million *other* galaxies in the universe. Regardless of how unusual our achievement might be, aren't we humans, after all, a fundamentally natural phenomenon? This seems to be the underlying message of all our scientific insights of recent years. The epic of evolution as it now stands has no major lacunae. Almost all the data contribute to a single coherent picture, and we fit right in. Dinosaurs did not live in the coal swamps of the Carboniferous period, larger mammals did not share the environment in which dinosaurs reigned supreme, and all hominid bones that can be dated are no more than a few million years old. Physicists, estimating time by measuring isotopic ratios, provide a concordant and reasonable chronology. All told, it's a consistent and seamless story.[42]

When viewed from the perspective of this larger odyssey, supported by a wealth of continually accumulating knowledge, one can only surmise that we humans are very rare and very special. Rare, because the requirements for life, especially intelligent terrestrial life, are such an uncommon and slender sliver of the wide array of physical parameters to be found in the universe at large. Humankind is special because of the many factors and contingencies that have fostered our early history, and propelled us to ever higher levels of cognition. Yet, shouldn't we expect nature, in the immensity of time and space, to have created similar odysseys elsewhere? From a scientific perspective, vast distances and unimaginably great periods of time demolish the likelihood that we will ever verify the existence of those other intelligent creatures beyond our star. Such isolation, within our current understanding of the true dimensions of the universe, may be reason enough to feel alone, but why should we feel insignificant?

"Man carries the world in his head, the whole astronomy and chemistry suspended in a thought. Because the history of nature is charactered in his brain, therefore is he the prophet and discoverer of her secrets," claimed Ralph Waldo Emerson.[43]

Human cognitive abilities are nothing less than awesome. But such cognition can also vividly experience dangers, both real and imagined, immediately or far into the future. Such a mind becomes frightened by its frailty and apprehensive of an uncertain fate. All human societies have responded to this dilemma by creating religious traditions and philosophies to strengthen our resolve and sustain our hope. Philosophies and religions have created mythical epics and a body of ethics in which humankind and the world of nature find their meaning and significance.

In contrast, science is only a clever methodology that enables us to peer into the nature of things. Tightly constrained by its restriction to physical data, scientific methodology has, nevertheless, created a panoramic vision of extraordinary depth and unity. "The universe we live in is beautiful," declares cosmologist Lee Smolin, ". . . because a multitude of phenomena are taking place on a vast array of different scales."[44] At every level where we have been able to probe, from deep within the interior of the atom to the clustering of galaxies, we have found complex structure and dynamic interactions.[45] This complexity has a hierarchial organization, in which higher levels often introduce radically new parameters not clearly determined by the levels beneath.

Though lacking the purpose and meaning people find so satisfying in philosophy and religious experience, the scientific enterprise has presented us with a hugely awesome firmament. This grand universe has itself grown more complex and richer with possibilities over an immense progression of time. In such a universe, it seems likely that creatures with higher cognitive intelligence, like shooting stars that suddenly flash across the black vault of night, come into being from time to time, then quickly fade away, surviving only as the stardust from which they were created.

We may never know those other twinklings of intelligence, rare and distant over so vast an expanse of space and time. Yet let us rejoice that we ourselves, if only for a brief moment, have become one such flash of cosmic understanding.

# NOTES

clever species

## INTRODUCTION

1. Claiming that the Atlantic has evolved is, perhaps, a generous use of the verb *to evolve*, but it is consistent with common modern usage. Darwin rarely used the word *evolution*, preferring instead the phrase *descent with modification*. In this way he avoided both the "unfolding" of an already formed plan, which was what earlier usage of the word *evolution* implied, and also the notion of an upward progression.

2. Compared with Freud's, Darwin's ideas had relatively little impact on the more elite aspects of our intellectual life in the twentieth century; see for example Richard Tarnas, *The Passion of the Western Mind* (New York: Ballantine Books, 1991). Molecular biology is helping change this; for a short summary of Darwin's impact see Ernst Mayr, "Darwin's Influence on Modern Thought," *Scientific American* 283, no.1 (July 2000): 79–83.

# 1. NO ORDINARY STAR

1. The conviction that our galaxy should be rich in technological civilizations inspired the search for extraterrestrial intelligence (SETI). For references, see note 7 in chapter 11.

2. John Horgan's book *The End of Science* (Reading, Mass.: Addison-Wesley, 1996) claims that science has just about run out of fundamental things to explore. However, as regards the physical sciences, Neil de Grasse Tyson demolishes Horgan's thesis in a just a few pages; see "The Beginning of Science," *Natural History* 110, no. 1 (February 2001): 80–83.

3. For general discussions regarding our galaxy and cosmic origins, see Ken Croswell, *The Alchemy of the Heavens* (New York: Anchor Press 1995); Armand Delsemme, *Our Cosmic Origins* (Cambridge, England: Cambridge University Press, 1998), and Lee Smolin, *The Life of the Cosmos* (New York: Oxford University Press, 1997). For a short review covering basic physics, chemistry, and astronomy, see John Gribbin, *Almost Everyone's Guide to Science: The Universe, Life, and Everything* (New Haven, Conn.: Yale University Press, 1999).

4. Croswell, *The Alchemy of the Heavens*.

5. Priscilla Frisch, "The Galactic Environment of the Sun," *American Scientist* 88 (1988): 52–59.

6. Andrew Clark and David Clark, *Aliens: Can We Make Contact with Extraterrestrial Intelligence?* (New York: Fromm International, 1999), p. 105.

7. For a fine review of the origin of the elements and their relative abundance, see Marcus Chown's *The Magic Furnace: The Search for the Origin of Atoms* (New York: Oxford University Press, 2001).

8. T. Snow and A. Witt, "The Interstellar Carbon Budget and the Role of Carbon in Dust and Large Molecules," *Science* 270 (1995): 1455–60.

9. See the report by P. Hoppe et al.,"Type II Supernova Matter in a Silicon Carbide Grain from the Murchison Meteorite," *Science* 272 (1996): 1314–16.

10. Delsemme, *Our Cosmic Origins*, p. 78.

11. For a recent review of the Sun's structure and its effects on our atmosphere, see Leon Golub and Jay Pasachoff's *Nearest Star: The Surprising Science of Our Sun* (Cambridge, Mass.: Harvard University Press, 2001).

12. Rudolf Kippenhahn, *100 Billion Suns: The Birth, Life, and Death of Stars* (New York: Basic Books, 1983), p. 237.

13. D. Black, "Completing the Copernican Revolution: The Search for Other Planetary Systems," *Annual Review of Astronomy and Astrophysics* 33 (1995): 359–80.

14. *Time* (5 February 1996). For a recent listing of the extrasolar planets, see

David Darling, *The Extraterrestrial Encyclopedia* (New York: Three Rivers Press, 2000), p. 137.

15. These techniques and the astronomers who developed them are described in Michael Lemonick's *Other Worlds: The Search for Life in the Universe* (New York: Simon and Schuster, 1998) and Ken Croswell's *Planet Quest: The Epic Discovery of Alien Solar Systems* (New York: The Free Press, 1997). For a more technical review, see G. W. Marcy and R. P. Butler, "Detection of Extrasolar Giant Planets," *Annual Review of Astronomy and Astrophysics* 36 (1998): 57–97. Short reviews are found in chapter 5 of David Koerner and Simon LeVay's *Here Be Dragons* (New York: Oxford University Press, 2000) and David Darling's *The Extraterrestrial Encyclopedia* (New York: Three Rivers Press, 2000), p. 351.

16. Many astronomers place little value on the Titius-Bode rule. For the status of Pluto, see Neil de Grasse Tyson, "Pluto's Honor," *Natural History* 108, no. 1 (February 1999): 82–84, and Gribbin, *Almost Everyone's Guide to Science*, p. 162.

17. Lemonick, *Other Worlds*, p. 182.

18. Peter Ward and Donald Brownlee, *Rare Earth: Why Complex Life Is Uncommon in the Universe* (New York: Copernicus Books, 2000).

19. Ken Croswell, "Why Intelligent Life Needs Giant Planets," *New Scientist* 136, no. 1844 (1992), p. 18. See also Ward and Brownlee's *Rare Earth*.

## 2. A REALLY GOOD PLANET IS HARD TO FIND

1. For a general background on molecular and atomic structure in an accessible text, see chapters 4 and 5 in John Gribbin's *Almost Everyone's Guide to Science: The Universe, Life, and Everything* (New Haven, Conn.: Yale University Press, 1999).

2. Peter Ward and Donald Brownlee discuss many aspects of "habitable zones" in *Rare Earth: Why Complex Life Is Uncommon in the Universe* (New York: Copernicus Books, 2000).

3. This change in nearness to the Sun progresses through the year in a cycle of about 40,000 years and has been a factor in recent glacial episodes.

4. M. Rampino and K. Caldera, "The Goldilocks Problem: Climatic Evolution and Long-Term Habitability of Terrestrial Planets," *Annual Review of Astronomy and Astrophysics* 32 (1994): 83–114.

5. Neil Comins, *What If the Moon Didn't Exist?* (New York: HarperCollins, 1993).

6. A. N. Halliday and M. J. Drake, "Colliding Theories," *Science* 283 (1999): 1861–62; J. L. Lissauer, "It's Not Easy to Make the Moon," *Nature* 389 (1997): 327–28.

7. Alfred Wegener, *The Origins of Continents and Oceans*. Translated from the fourth revised German edition by John Birham. (New York: Dover, 1966).

8. William Glen gives a technical account of how a variety of independent research programs contributed to a final synthesis in *The Road to Jaramillo: Critical Years of the Revolution in Earth Science* (Stanford, Calif.: Stanford University Press, 1982). A more general discussion is found in Tjeerd Van Andel's *New Views on an Old Planet: A History of Global Change*, 2d ed. (Cambridge, England: Cambridge University Press, 1994). Naomi Oreskes provides a detailed analysis of why North American geologists were so reluctant to accept Wegener's theory in *The Rejection of Continental Drift: Theory and Practice in American Earth Science* (New York: Oxford University Press, 1999).

9. If we map the alternating north- and south-pointing geomagnetic rocks by depicting one black and the other white, we find symmetric, zebralike patterns paralleling the deep ocean ridges. Geological maps of continental surfaces do not exhibit similar patterns.

10. S. R. Taylor and S. McLennan, "The Evolution of the Continental Crust," *Scientific American* 274, no. 1 (January 1996): 76–81.

11. Recently discovered hot springs, deep on the ocean floor between New Guinea and New Ireland, are rich in concentrated metals; see W. J. Broad, "First Move Is Made to Mine the Seabed," *New York Times* (1 December 1997): 1.

12. Ward and Brownlee discuss many additional aspects of plate tectonics and claim that it is central to maintaining climate stability over millions of years in *Rare Earth*. Other recent books that discuss aspects of our planet and its history in greater detail are J. D. Macdougal, *A Short History of Planet Earth* (New York: John Wiley and Sons, 1996); and Harry McSween Jr., *Fanfare for Earth: The Origin of Our Planet and Life* (New York: St. Martin's Press, 1997); Stuart R. Taylor, *Destiny or Chance: Our Solar System and its Place in the Cosmos* (Cambridge, England: Cambridge University Press, 1998).

13. Armand Delsemme, *Our Cosmic Origins* (Cambridge, England: Cambridge University Press, 1998).

14. David Morrison, *Exploring Planetary Worlds* (New York: Scientific American Library, 1993).

15. Stuart R. Taylor, *Destiny or Chance: Our Solar System and Its Place in the Cosmos* (Cambridge, England: Cambridge University Press, 1998), p. 202.

16. The Earth may in fact have suffered from episodes of global frigidity and frozen oceans during its first few thousand million years, see: Kerr, "An

Appealing Snowball Earth That's Still Hard to Swallow," *Science* 287 (2000): 1734–36; J. L. Kirschvink et al., "Paleoproterozoic Snowball Earth: Extreme Climate and Geochemical Global Change and Its Biological Consequences," *Proceedings of the National Academy of Sciences USA* 97 (2000): 1400–1405.

17. R. A. Berner, "The Carbon Cycle and $CO_2$ over Phanerozoic Time: The Role of Land Plants," *Philosophical Transactions of the Royal Society, London* B 353 (1998): 75–82.

18. Other suggestions for cooling over the last fifty million years can be found in R. C. Wilson, S. A. Drury, and J. L. Chapman, *The Great Ice Age: Climate Change and Life* (London, Routledge, 2000).

19. Stuart Taylor comes to much the same conclusion: "The message is that chance events have played a crucial role in the origin and evolution of the solar system, and that planets similar to earth are unlikely in other planetary systems," *Destiny or Chance: Our Solar System and Its Place in the Cosmos* (Cambridge, England: Cambridge University Press, 1998), p. 134. An even more persuasive case is made by Ward and Brownlee in *Rare Earth.*

## 3. THE FIRST 4,000 MILLION YEARS

1. C. J. Bult, et al., "Complete Genome Sequence of the Methanogenic Archaeon, *Methanococcus jannaschii*," *Science* 273 (1996): 1058–72.

2. E. Pennisi, "Is It Time to Uproot the Tree of Life?" *Science* 284 (1999): 1305–1307.

3. Kauffman's ideas, while quite abstract, are enthusiastically presented in his *At Home in the Universe* (New York: Oxford University Press, 1995). These ideas are part of a more general discussion in M. Mitchel Waldrop's *Complexity: The Emerging Science at the Edge of Order and Chaos* (New York: Simon and Schuster, 1992).

4. W. A. Bentley and W. J. Humphries, *Snow Crystals* (New York: Dover, 1962).

5. This claim was made despite evidence that the spherules were imbedded in a mineral matrix probably formed at temperatures above 500°C. See also R. Kerr, "Putative Martian Microbes Called Microscopy Artifacts," *Science* 278 (1998): 1706–1707.

6. David Koerner and Simon LeVay are more sympathetic to the search for evidence of life in this Martian rock; see *Here Be Dragons* (New York: Oxford University Press, 2000), pp. 75–81.

7. Stuart Kauffman as quoted in Waldrop's *Complexity*, p. 321.

8. Per Bak, *How Nature Works: The Science of Self-Organized Criticality* (New York: Springer-Verlag New York: 1996).

9. Christian de Duve discusses these ideas with clarity and depth in *Vital Dust: Life as a Cosmic Imperative* (New York: Basic Books, 1995).

10. Further general texts on the origins of life for readers with a better background in biology are John Maynard Smith and Eörs Szathmary's *The Major Transitions in Evolution* (New York: W. H. Freeman, 1995) and their more abbreviated *The Origins of Life: From the Birth of Life to the Origins of Language* (New York: Oxford University Press, 1999).

11. H. D. Holland, "Evidence for Life on Earth More Than 3,850 Million Years Ago," *Science* 275 (1997): 38–39.

12. Lynn Margulis and Karlene Schwartz, *Five Kingdoms: An Illustrated Guide to the Phyla of Life on Earth* (New York: W. H. Freeman, 1982), p. 46.

13. J. William Schopf, "The Oldest Fossils and What They Mean," in J. *Major Events in the History of Life*, ed. William Schopf (Boston: Jones and Bartlett, 1992), pp. 29–63.

14. T-S. Han and B. Runnegar, "Megascopic Eukaryotic Algae from the 2.1 Billion-Year-Old Negaunee Iron-Formation, Michigan," *Science* 257 (1992): 232–35.

15. The many intricate activities of the cell are clearly described in Boyce Rensberger's *Life Itself: Exploring the Realm of the Living Cell* (New York: Oxford University Press, 1996). Not as up to date, but with fundamental insights into the nature of cellular biochemistry, are chapters 3 through 5 of Jacques Monod's classic *Chance and Necessity: An Essay on the Natural Philosophy of Modern Biology* (New York: Alfred A. Knopf, 1971).

16. K. Nasmyth, "Putting the Cell Cycle in Order," *Science* 274 (1996): 1643–45.

17. Lynn Margulis, *Origin of Eukaryotic Cells* (New Haven, Conn.: Yale University Press, 1970).

18. William Paley, *Natural Theology: Or Evidences of the Existence and Attributes of the Diety, Collected from the Appearances of Nature*. Reprint of the 12th ed., 1809. (1809; reprint, Charlottesville, Va.: Ibis Publications, 1986).

19. C. F. Delwiche, "Tracing the Thread of Plastid Diversity through the Tapestry of Life," *American Naturalist* 154 (1999): S164–77.

20. T. Vellai et al., "A New Aspect to the Origin and Evolution of Eukaryotes," *Journal of Molecular Evolution* 46 (1998): 499–507. See also T. Vellai and G. Vida, "The Origin of Eukaryotes: The Difference between Prokaryotic and Eukaryotic Cells," *Proceedings of the Royal Society, London* B 266 (1999): 1571–77.

21. John Maynard Smith and Eörs Szathmary suggest that the loss of the rigid outer bacterial cell wall required the larger cell to develop a more robust internal architecture; see *The Origins of Life*, p. 62.

22. Even the malaria parasite has acquired an unusual symbiont and this knowledge may help us fight the parasite; H. Jommaa et al., "Inhibitors of the Non-mevalonate Pathway of Isoprenoid Biosynthesis as Antimalarial Drugs," *Science* 285 (1999): 1573–76.

23. Some British authors state Murphy's Law as "What can happen, will happen." The American version (going back to a U.S. Air Force engineer named Murphy) is used here.

24. A few authors have argued that the genes themselves are the primary determinants of evolution. Not so; it is the individual organism (or group of cooperating organisms) that must produce vigorous offspring, able to carry their genes forward through future generations. Parasitic genes may hang on within the genome and be carried forward through time, but if they become too great a burden the organism will fail.

25. In fact, genes are not the simple entities we once envisioned. While most are clearly separated along the chromosome, others overlap, and some even jump around.

26. Chimpanzee DNA is estimated to differ from ours by between 1 and 2 percent—not a whole lot. But the great apes do have twenty-four pairs of chromsomes in contrast to our twenty-three pairs. See M. Ruvola, "Molecular Phylogeny of the Hominoids: Inferences from Multiple Independent DNA Sequence Data Sets," *Molecular Biology and Evolution* 14 (1997): 248–65.

27. For more detailed discussions regarding the reasons for sexual reproduction, see Smith and Szathmary's *The Origins of Life* and George Williams, *The Pony Fish's Glow and Other Clues to Plan and Purpose in Nature* (New York: Basic Books, 1997).

28. Keeping the life span short and population turnover rapid is another way to thwart pathogens. For a short review of aging and the factors responsible for life spans, see T. B. Kirkwood and S. N. Austad, "Why Do We Age?" *Nature* 408 (2000): 233–38.

29. Fred Hoyle and Chandra Wickramasinghe, 1993. *Our Place in the Cosmos: The Unfinished Revolution* (London: J. M. Dart, 1993).

30. Michael J. Behe, *Darwin's Black Box: The Biochemical Challenge to Evolution* (New York: Free Press, 1996). For a strong scientific rejoinder to claims of intelligent design and many other aspects of "the new creationism," see Robert Pennock's *Tower of Babel: The Evidence against the New Creationism* (Cambridge, Mass.: MIT Press, 1999).

31. William Dembski, another advocate of divine design, calls this "promissory materialism," but it's a strategy that has worked remarkably well; see *Intelligent Design: The Bridge between Science and Theology* (Downers Grove, Ill.: Intervarsity Press, 1999) p. 218.

32. The earliest evidence for what may be sex cells in an alga have recently been found in rocks dated at 1,200 million years old; see N. J. Butterfield, "*Bangiomorpha pubescens* n. gen., n. sp.; Implication for the Evolution of Sex, Multicellularity, and the Mesoproterozoic/Neoproterozoic Radiation of Eukaryotes," *Paleobiology* 26 (2000): 386–404.

33. Peter Ward and Donald Brownlee discuss many aspects of "snowball Earth" scenarios in chapter 6 of their book *Rare Earth: Why Complex Life Is Uncommon in the Universe* (New York: Copernicus Books, 2000).

## 4. LIFE EXPLODES: THE LAST 560 MILLION YEARS

1. M. A. Fedonkin, "Cold Water Origin of Animal Life," Lecture at Field Museum, Chicago, 10 Dec. 1996.

2. M. A. Fedonkin and B. M. Waggoner, "The Late Precambrian Fossil *Kimberella* is a Mollusc-like Bilaterian Organism," *Nature* 388 (1997): 868–71.

3. S. Bengtson, "Animal Embryos in Deep Time," *Nature* 391 (1998): 529–30; S. Xiao, Y. Zhang and A. H. Knoll, "Three-dimensional Preservation of Algae and Animal Embryos in a Neoproterozoic Phosphorite," *Nature* 391 (1998): 553–58.

4. Richard Fortey, *Life: A Natural History of the First Four Billion Years of Life on Earth* (New York: Alfred A. Knopf, 1998), p. 10.

5. Many of these ancient fossil localities do not have precisely datable associated strata, so that the actual time periods may be approximations.

6. E. H. Davidson, K. J. Peterson, and R. A. Cameron, "Origin of Bilaterian Body Plans: Evolution of Developmental Regulatory Mechanisms," *Science* 270 (1995): 1319–25.

7. F. J. Ayala, A. Rzhetsky, and F. J. Ayala, "Origin of the Metazoan Phyla: Molecular Clocks Confirm Paleontological Estimates," *Proceedings of the National Academy of Sciences USA* 95 (1998): 606–11; see also: D. Erwin, J. Valentine, and D. Jablonski, "The Origin of Animal Body Plans," *American Scientist* 85 (1997): 126–37.

8. Cells and cell functions are also more complex than we thought. See Paul Nurse, "The Incredible Life and Times of Biological Cells," *Science* 289 (200): 1711–16.

9. S. B. Carroll, "Chance and Necessity: The Evolution of Morphological Complexity and Diversity," *Nature* 409 (2001): 1102–1109.

10. For further discussion, see Rudolf Raff, *The Shape of Life: Genes, Development, and the Evolution of Animal Form* (Chicago: University of Chicago Press, 1996).

11. A. H. Knoll, "The Early Evolution of Eukaryotes: A Geological Perspective," *Science* 256 (1992): 622–27 and A. H. Knoll and S. B. Carroll, "Early Animal Evolution: Emerging Views from Comparative Biology and Geology," *Science* 284 (1999): 2129–37.

12. Stephen J. Gould, *Wonderful Life: The Burgess Shale and the Nature of History* (New York: W. W. Norton, 1989).

13. Simon Conway Morris, *The Crucible of Creation: The Burgess Shale and the Rise of Animals* (New York: Oxford University Press, 1998).

14. For a detailed discussion with a multitude of references regarding catastrophism, see Trevor Palmer's *Controversy, Catastrophism, and Evolution: The Ongoing Debate* (New York: Kluwer Plenum, 1999). But be warned, Palmer gives some very weird ideas far more consideration than they deserve.

15. Important references are L. W. Alvarez et al., "Extraterrestrial Cause for the Cretaceous-Tertiary Extinction," *Science* 208 (1980): 1095–1108; Walter Alvarez, *T. Rex and the Crater of Doom* (Princeton, N.J.: Princeton University Press, 1997); and William Glen, ed. *The Mass Extinction Debate: How Science Works in a Crisis* (Stanford, Calif.: Stanford University Press, 1994).

16. The terminal Cretaceous impact structure is one of three such large features known on the Earth's surface.

17. A few scholars still dispute this conclusion. The end of the Cretaceous was also the time interval during which huge outpourings of lava formed the Deccan flood basalts in India; this massive vulcanism may have been an additional contributing factor to the extinction.

18. In a lucid discussion of extinctions, David Raup claims that sudden extinctions have played a significant role in promoting diversity by keeping "the pot well stirred," see *Extinction: Bad Genes or Bad Luck*? (New York: W. W. Norton, 1991), p. 20. In contrast, Peter Ward and Donald Brownlee argue that asteroidal impacts are a major threat to complex life forms in *Rare Earth: Why Complex Life Is Uncommon in the Universe* (New York: Copernicus Books, 2000), p. 181.

19. The early parts of this story are nicely described by David Raup in *The Nemesis Affair: A Story of the Death of Dinosaurs and the Ways of Science* (New York: W. W. Norton, 1986). The latest suggestion is that our solar system is perturbed as it oscillates "up and down" through the galactic plane while it circles the

galactic center; see M. R. Rampino, "Impact Crises, Mass Extinctions, and Galactic Dynamics: The Case for a Unified Theory," in "Large Meteorite Impacts and Planetary Evolution II: Boulder, Colorado," B. O. Dressler and V. L. Sharpton, eds., *Geological Society of America Special Paper* No. 339 (1999), pp. 241–48.

20. P. Kenrick and P. R. Crane, "Origin and Early Evolution of Plants on Land," *Nature* 389 (1997): 33–34; J. B. Richardson, "Origin and Evolution of the Earliest Land Plants," in *Major Events in the History of Life*, ed. J. William Schopf (Boston: Jones and Bartlett, 1992), pp. 95–118.

21. Such a long period of ice ages contradicts the more enthusiastic promoters of the Gaia hypothesis, who claim that the Earth and its biosphere together form a grand self-correcting and self-stabilizing system. For more about the Gaia hypothesis, see chapter 8 in Lynn Margulis, *Symbiotic Planet: A New View of Evolution* (New York: Basic Books, 1998) and Elisabet Sahtouris, *Gaia: The Human Journey from Chaos to Cosmos* (New York: Pocket Books, 1989). For short reviews see C. Barlow and T. Volk, "Gaia and Evolutionary Biology," *Bioscience* (1992): 686–92, and S. Schneider, "A Goddess of Earth or the Imagination of a Man?" *Science* 291 (2001): 1906–1907.

22. Niles Eldredge, *Life Pulse: Episodes from the Story of the Fossil Record* (New York: Facts on File, 1987).

23. M. Hasebe and M. Ito, "Evolution of Reproductive Organs in Vascular Plants," in *The Biology of Biodiversity*, M. Kato, ed. (Tokyo: Springer Verlag, 1999), pp. 243–55.

24. J. J. Sepkoski, "A Factor Analytic Description of the Phanerozoic Marine Fossil Record," *Paleobiology* 7 (1981): 36–53.

## 5. WHY ARE THERE SO MANY KINDS OF PLANTS AND ANIMALS?

1. M. J. Donoghue and W. S. Alverson, "A New Age of Discovery," *Annals of the Missouri Botanical Garden* 87 (2000): 110–26.

2. J. C. Briggs, "Species Diversity: Land and Sea Compared," *Systematic Biology* 43 (1994): 130–35.

3. In contrast, very cold and deep marine environments may have achieved their present diversity only recently.

4. A few authors claim that the microscopic life of the sea is grossly underestimated and that there may be somewhere between 100,000 and 10,000,000 species of diatoms alone; see T. A. Norton, M. Melkonian, and R. A. Andersen, "Algal Biodiversity," *Phycologia* 35 (1996): 308–26.

5. W. C. Burger, "Why Are There So Many Kinds of Flowering Plants in Costa Rica?" in *The Botany and Natural History of Panama*, ed. William D'Arcy and Mireya Correa (St. Louis: Botanical Garden, 1980), pp. 125–36.

6. A. Henderson, S. P. Churchill, and J. L. Luteyn, "Neotropical Plant Diversity," *Nature* 351 (1991): 21–22.

7. Robin Foster, personal communication (May 1999).

8. S. R. Hubbell and R. B. Foster, "Biology, Chance, and History and the Structure of Tropical Rain Forest Tree Communities," in *Community Ecology*, ed. Jared Diamond and T. J. Case (New York: Harper and Row, 1986), pp. 314–19.

9. In Simon Levin's phrasing: "Diversity is maintained because of the global certainty of local uncertainty; that is, the continual renewal of opportunity through localized and random disturbances," *Fragile Dominion: Complexity and the Commons* (Reading, Mass.: Perseus Press, 1999), p. 113.

10. W. C. Burger, "Why Are There So Many Kinds of Flowering Plants?" *Bioscience* 31 (1981): 572–81.

11. T. Southwood, "The Number of Species of Insect Associated with Various Trees," *Journal of Animal Ecology* 30 (1961): 1–8.

12. Edward O. Wilson, *The Diversity of Life* (Cambridge, Mass.: Harvard University Press, 1992), p. 204.

13. J. U. Ganzhorn et al., "Habitat Characteristics and Lemur Species Richness in Madagascar," *Biotropica* 29 (1997): 331–43.

14. G. Oba et al., "New Perspectives on Sustainable Grazing Management in Arid Zones of Sub-Saharan Africa," *Bioscience* 50 (2000): 35–51.

15. For an introduction to floral biology, see Peter Bernhardt's *The Rose's Kiss: A Natural History of Flowers* (Washington, D.C.: Island Press, 1999).

16. Also, it appears that beetles have increased their numbers *because of* flowering plants; see B. D. Farrell, " 'Inordinate Fondness' Explained. Why Are There So Many Beetles?" *Science* 281 (1998): 555–59.

17. Ernst Mayr, *Animal Species and Evolution* (Cambridge, Mass.: Harvard University Press, 1963).

18. Charles Darwin, *The Origin of Species by Means of Natural Selection, or the Preservation of Favored Races in the Struggle for Life*, 6th ed. (New York: D. Appleton, 1895), p. 136.

19. K. E. Filchak et al., "Natural Selection and Sympatric Divergence in the Apple Maggot *Rhagoletis pomonei*," *Nature* 407 (2000): 739–42.

20. R. Tryon, "Development and Evolution of Fern Floras of Oceanic Islands," *Biotropica* 2 (1970): 76–84.

21. C. Schneider and C. Moritz, "Rainforest Refugia and Evolution in Aus-

tralia's Wet Tropics," *Proceedings of the Royal Society, London* B 266 (1999): 191–96.

22. D. G. Campbell, "Splendid Isolation in Thingvallavatan," *Natural History* 105, no. 6 (June 1996): 48–55. For an excellent discussion of speciation, see Menno Schilthuisen, *Frogs, Flies, and Dandelions* (Oxford, England: Oxford University Press, 2001).

23. This story is described from a very personal point of view in Tijs Goldschmidt, *Darwin's Dreampond; Drama in Lake Victoria* (Boston: MIT Press, 1996).

24. A more traditional explanation invoking microgeographical speciation is given by George Barlow in *The Cichlid Fishes: Nature's Grand Experiment in Evolution* (Cambridge, Mass.: Perseus, 2000).

25. O. Seehausen, J. van Alphen, and F. Witte, "Cichlid Fish Diversity Threatened by Eutrophication That Curbs Sexual Selection," *Science* 277 (1997): 1808–11.

26. The work on Darwin's finches by Rosemary and Peter Grant is engagingly described in Jonathan Weiner's *The Beak of the Finch; A Story of Evolution in our Time* (New York: Vintage Books, Random House, 1994).

27. What's interesting here is that some of the seeds of these close congeners must be distributed widely and into the each other's habitats, but they reach maturity only in their own specific habitat. See W. C. Burger, "Montane Species-Limits in Costa Rica and Evidence for Local Speciation on Altitudinal Gradients," in *Biodiversity and Conservation of Neotropical Montane Forests*, ed. S. P. Churchill et al. (New York: Botanical Garden, 1995), pp. 127–33.

28. For a discussion of Wallace's suggestion, see Verne Grant, *Plant Speciation* (New York: Columbia University Press, 1971), pp. 135–49.

29. With his characteristic clarity and verve, Stephen Jay Gould explains why the Linnaean system, developed under the assumptions of theistic creationism, has continued to function so well in a world of Darwinian evolution; see "Linnaeus's Luck?" *Natural History* 109, no. 7 (September 2000): 18–25, 66–76.

30. J. Sepkoski et al., "Phanerozoic Marine Diversity and the Fossil Record," *Nature* 293 (1981): 435–37.

31. S. W. Wilhelm and C. A. Suttle, "Viruses and Nutrient Cycles in the Sea," *Bioscience* 49 (1999): 781–88.

32. S. J. Gould, "On Replacing the Idea of Progress with an Operational Notion of Directionality," in *Evolutionary Progress?* ed. Matthew H. Nitecki (Chicago: University of Chicago Press, 1988), pp. 319–38.

## 6. RANDOM WALKS OR PROGRESSIVE TRAJECTORIES?

1. Thomas Kuhn, *The Structure of Scientific Revolutions*, 2d ed. (Chicago: University of Chicago Press, 1970), p. 172.

2. Ernst Mayr, *The Growth of Biological Thought: Diversity, Evolution, and Inheritance* (Cambridge, Mass.: Harvard University Press, 1982), p. 532.

3. Julian Huxley, *Evolution in Action* (New York: Harper and Brothers, 1953), p. 126.

4. George Gaylord Simpson, *The Meaning of Evolution* (New Haven, Conn.: Yale University Press, 1949).

5. Niles Eldredge, *Life Pulse: Episodes from the Story of the Fossil Record* (New York: Facts on File, 1987).

6. M. Wills, "Crustacean Disparity through the Phanerozoic: Comparing Morphological and Stratigraphic Data," *Biological Journal of the Linnaean Society* 65 (1998): 455–500.

7. W. B. Saunders et al., "Evolution of Complexity in Paleozoic Ammonoid Sutures," *Science* 286 (1999): 760–63.

8. Michael Ruse, *Monad to Man: The Concept of Progress in Evolutionary Biology* (Cambridge, Mass.: Harvard University Press, 1996), p. 416.

9. Stephen J. Gould, *Full House: The Spread of Excellence from Plato to Darwin* (New York: Harmony Books, 1996), pp. 3, 20.

10. J. Alroy, "Cope's Rule and the Dynamics of Body Mass: Evolution in North American Fossil Mammals," *Science* 280 (1998): 731–34.

11. Jacob Bronowski, "New Concepts in the Evolution of Complexity," *Synthese* 21 (1970): 228–45.

12. Alison Jolly, *Lucy's Legacy: Sex and Intelligence in Human Evolution* (Cambridge, Mass.: Harvard University Press, 1999), p. 22.

13. Robert Wright argues vigorously for both cultural and biological progress; he insists that there is a forward-thrusting "kind of force" which results in "the playing of ever-more-elaborate zero-sum-games." See *Nonzero: The Logic of Human Destiny* (New York: Pantheon Books, 2000), p. 6.

14. G. Ledyard Stebbins, *The Basis of Progressive Evolution* (Chapel Hill: University of North Carolina Press, 1969), p. 142.

15. Richard Fortey, *Life: A Natural History of the First Four Billion Years of Life on Earth* (New York: Alfred A. Knopf, 1998), p. 168.

16. Ian Stewart and Jack Cohen, *Figments of Reality: The Evolution of the Curious Mind* (Cambridge, England: Cambridge University Press, 1997).

17. John T. Bonner, *The Evolution of Complexity by Means of Natural Selection* (Princeton, N.J.: Princeton University Press, 1988).

18. Michael Majerus, *Melanism: Evolution in Action* (New York: Oxford University Press, 1998).

19. A. M. Shapiro, "Egg-Mimics of *Strepanthus* (Cruciferae) Deter Oviposition by *Pieris sisymbrii* (Lepidoptera: Pieridae)," *Oecologia* 48 (1981): 142–45.

20. Our vermiform appendix, so prone to infection, has been thought to be a "vestigial organ" from earlier times, but it may still play a role in early development.

21. F. Schiestl et al., "Orchid Pollination by Sexual Swindle," *Nature* 399 (1999): 421–22.

22. Darwin's interpretation has recently been challenged by a hypothesis suggesting that a pollinator-shift mechanism may have been involved; see L. A. Nilsson et al., "Monophily and Pollination Mechanisms in *Angraecum arachnites* Schltr. (Orchidaceae) in a Guild of Long-Tongued Hawk-Moths (Sphingidae) in Madagascar," *Biological Journal of the Linnaean Society* 26 (1985): 1–19.

23. D. S. Seigler and J. Ebinger, "Cyanogenic Glycosides in Ant-Acacias of Mexico and Central America," *Southwest Naturalist* 32 (1987): 459–503.

24. B. D. Farrell, "'Inordinate Fondness' Explained. Why Are There So Many Beetles?" *Science* 281 (1998): 555–59.

25. J. B. Losos et al., "Contingency and Determinism in Replicated Adaptive Radiations of Island Lizards," *Science* 279 (1998): 2115–18.

26. Richard Dawkins, *The Blind Watchmaker: Why Evidence of Evolution Reveals a Universe without Design* (New York: W. W. Norton, 1985), p. 181.

27. Geerat J. Vermeij, *Evolution and Escalation: An Ecological History of Life* (Princeton, N.J.: Princeton University Press, 1987).

28. Michael W. Majerus, William Amos, and Gregory Hurst, *Evolution: The Four Billion Year War* (Essex, England: Longmans, 1996), p. 156.

29. For an excellent discussion of chemicals in the ecology and interactions of plants and animals, see William Agosta, *Bombardier Beetles and Fever Trees* (Reading, Mass.: Helix Books, 1996).

30. These competitive interactions may take on other characteristics as they extend over time and become more like "trench warfare" than progressive arms races; see E. A. Stahl et al., "Dynamics of Disease Resistance Polymorphism at the *Rpm1* Locus of *Arabidopsis*," *Nature* 400 (1999): 667–71.

31. R. Byrne and S. P. Horn, "Prehistoric Agriculture and Forest Clearance in the Sierra de Los Tuxtlas, Veracruz, Mexico," *Palynology* 13 (1989): 181–93.

32. Jared Diamond, *Guns, Germs, and Steel: The Fates of Human Societies* (New York: W. W. Norton, 1997) and Roy Porter, *The Greatest Benefit to Mankind: A Medical History of Humanity* (New York: W. W. Norton, 1998).

33. Leigh Van Valen, "A New Evolutionary Law," *Evolutionary Theory* 1 (1973): 1–30.

34. David M. Raup, "Testing the Fossil Record for Evolutionary Progress," in *Evolutionary Progress?* ed. M. Nitecki (Chicago, University of Chicago Press, 1980), pp. 293–317 and J. Sepkoski et al., "Phanerozoic Marine Diversity and the Fossil Record," *Nature* 293 (1981): 435–37.

35. Or might environmental disasters have diminished in frequency over the last several hundred million years? See M. Newman and G. Eble, "Decline in Extinction Rates and Scale Invariance in the Fossil Record," *Paleobiology* 25 (1999): 434–39.

36. Thomas Robert Malthus, *An Essay on the Principle of Population; or a View of Its Past and Present Effects* (London: Johnson, 1803).

37. David Quammen, *The Song of the Dodo: Island Biogeography in an Age of Extinctions* (New York: Scribner, 1996).

## 7. THE ESCALATION OF INTELLIGENCE

1. Harry Jerison, *Evolution of the Brain and Intelligence* (New York: Academic Press, 1973).

2. Irene Pepperberg documents the unusual cognitive abilities of parrots in *The Alex Studies: Cognitive and Communicative Abilities of Grey Parrots* (Cambridge, Mass.: Harvard University Press, 1999).

3. Harry Jerison, *Brain Size and the Evolution of Mind* (New York: American Museum of Natural History, 1991).

4. John Allman, *Evolving Brains* (New York: Scientific American Library, 1999).

5. This quarter-power scaling law appears to be due to the geometry of networks, such as blood vessels and capillaries, that are necessary to sustain complex organisms.

6. R. W. Sussman, "Primate Origins and the Evolution of Angiosperms," *American Journal of Primatology* 23 (1991): 209–23; "How Primates Invented the Rainforest and Vice Versa," in *Creatures of the Dark: The Nocturnal Prosimians*, ed. L. Alterman et al. (New York: Plenum Press, 1995), pp. 1–9.

7. Donald Johanson and Maitland Edey, *Lucy: The Beginnings of Humankind* (New York: Simon and Schuster, 1981). The earliest Ethiopian fragments, from before Lucy's time, are now being placed in the genus *Ardipithecus*; for a recent find, see *Nature* 412 (2001): 175–81, and commentary on pages 131–32.

8. For an in-depth review of the unique versatility of our hands and their importance in human evolution and behavior, see Frank R. Wilson's *The Hand: How Its Use Shapes the Brain, Language, and Human Culture* (New York: Pantheon Press, 1998).

9. This represents my view as a "lumper" who prefers to see these variable fossils as representing a single, long-lived, and quite variable lineage.

10. Yves Coppens, "East Side Story: the Origin of Humankind," *Scientific American* 270, no. 5 (May 1994): 88–95.

11. L. A. Isbell and T. P. Young, "The Evolution of Bipedalism in Hominids and Reduced Group Size in Chimpanzees; Alternative Responses to Decreasing Resource Availability," *Journal of Human Evolution* 30 (1996): 389–97.

12. Dean Falk, "Brain Evolution in *Homo*: The 'Radiator' Theory," *Behavioural and Brain Science* 13 (1990): 331–81.

13. S. M. Stanley, "An Ecological Theory for the Origin of *Homo*," *Paleobiology* 18 (1992): 237–57.

14. Jerison, *Brain Size and the Evolution of Mind.*

15. Interesting discussions of human intelligence and consciousness are available in Daniel Dennett's *Kinds of Minds: Toward an Understanding of Consciousness* (New York: Basic Books, 1996); Derek Denton, *The Pinnacle of Life: Consciousness and Self-Awareness in Humans and Animals* (New York: HarperCollins, 1993); and James Trefil, *Are We Unique? A Scientist Explores the Unparalleled Intelligence of the Human Mind* (New York: John Wiley, 1997). Francis Crick focuses on vision as a way of beginning to unravel the mystery of the human mind in *The Astonishing Hypothesis: The Scientific Search for the Soul* (New York: Scribners, 1994). Although a terrific book in many respects, A. G. Cairn-Smith's *Evolving the Mind: On the Nature of Matter and the Origin of Consciousness* (Cambridge, England: Cambridge University Press, 1996) concludes with the almost mystical suggestion that consciousness represents the manifestation of some as-yet-undiscovered macroquantum effect. Alwyn Scott dismisses this quantum effects idea, and makes a strong case for the "emergent" nature of human self-consciousness in *Stairway to the Mind: The Controversial New Science of Consciousness* (New York: Copernicus, 1995). In *The Private Life of the Brain: Emotions, Consciousness and the Secret of the Self* (New York: John Wiley, 2000), Susan Greenfield argues that emotions are the building blocks for developing consciousness, both in children and in our history. An interesting collection of essays, edited by Arnold Sheibel and J. William Schopf can be found in *The Origin and Evolution of Intelligence* (Sudbury, Mass.: Jones and Bartlett, 1997).

16. Despite their similarities, bonobos and chimpanzees exhibit great diversity in their DNA; see H. Kaessmann, V. Wiebe, and S. Pääbo, "Extensive

Nuclear DNA Sequence Diversity among Chimpanzees," *Science* 286 (1999): 1159–62.

17. Frans de Waal, *Good Natured: The Origin of Right and Wrong in Humans and Other Animals* (Cambridge, Mass.: Harvard University Press, 1996).

18. Much of the phrasing used here in discussing animal behavior is quite out of line with the mechanistic jargon so fashionable in the literature of modern behavioral science. Eileen Crist has contrasted the descriptive language of Darwin and French entomologist Fabre with that of modern behaviorists and concludes that "the conceptual ecologies created through technical [behaviorist] idioms are hostile to inner life, for such idioms tend to banish mentality either by force or by obviation. . . ." And she notes that "Darwin's anthropomorphic language emanates from a commitment to evolutionary continuity that, for him, inexorably includes behavioral and mental continuity." (*Images of Animals: Anthropomorphism and Animal Mind* [Philadelphia: Temple University Press, 1999], pp. 41, 124). I agree. Taking a very different position is Stephen Budiansky. In *If a Lion Could Talk* (New York: Free Press, 1998), he denigrates many aspects of animal cognition, including the story of Binti Jua. Compare Budiansky's negative attitude with Marc Bekoff's short review, "Animal Emotions: Exploring Passionate Natures," *Bioscience* 50 (2000): 861–67, or Donald Griffin's more comprehensive *Animal Minds* (Chicago: University of Chicago Press, 1992), and especially Frans de Waal, *The Ape and the Sushi Master* (New York: Basic Books, 2001).

19. See articles in *Self-Awareness in Animals and Humans*, ed. Sue Taylor Parker et al. (Cambridge, England: Cambridge University Press, 1994).

20. Jerison, *Brain Size and the Evolution of Mind*.

21. Robert McCormick Adams, *Paths of Fire: An Anthroplogist's Inquiry into Western Technology* (Princeton, N.J.: Princeton University Press, 1996), p. 36.

22. Robin Dunbar, *Grooming, Gossip, and the Evolution of Language* (Cambridge, Mass.: Harvard University Press, 1996).

23. A clever experiment with human infants and tamarin monkeys found that both could discriminate between sentences spoken in Dutch or in Japanese—but not if the sentences were played backward. Here's evidence for underlying auditory processes that go far back in primate history; see F. Ramus et al., "Language Discrimination by Human Newborns and Cotton-Top Tamarin Monkeys," *Science* 288 (2000): 349–51.

24. The gradual emergence of language out of earlier gestural forms of communication is convincingly argued in the book by David Armstrong, William Stokoe and Sherman Wilcox, *Gesture and the Nature of Language* (Cambridge, England: Cambridge University Press, 1995). See also M. C. Corballis, "The Gestural Origin of Language," *American Scientist* 87 (1999): 138–45.

25. Merlin Donald, *Origins of the Modern Mind* (Cambridge, Mass.: Harvard University Press, 1991).

26. Ibid.; Philip Lieberman, *The Biology and Evolution of Language* (Cambridge, Mass.: Harvard University Press, 1984); Ian Tattersall, *Becoming Human: Evolution and Human Uniqueness* (New York: Harcourt Brace, 1998).

27. P. V. Tobias, "Major Events in the History of Mankind," in *Major Events in the History of Life*, ed. J. William Schopf (Boston: Jones and Bartlett, 1992), pp. 141–75.

28. William Dembski claims that language is a divine gift but gives us no clue as to how this suggestion might be further evaluated; see *Intelligent Design: The Bridge between Science and Technology* (Downers Grove, Ill.: Intervarsity Press, 1991). Though focused on design, this book fails to mention one of nature's worst design blunders, the narrow pelvic opening through which human females are forced to deliver their big-brained babies.

29. R. Dennell, "The World's Oldest Spears," *Nature* 385 (1997): 767–68.

30. Steven Pinker, *The Language Instinct* (London: W. Morrow, 1994), p. 125.

31. Robert D. Martin, *Primate Origins and Evolution: A Phylogenetic Reconstruction* (London: Chapman and Hill, 1990), p. 391.

32. J. Kappelman, "The Evolution of Body Mass and Relative Brain Size in Fossil Hominids," *Journal of Human Evolution* 30 (1996): 243–76.

33. C. B. Ruff, E. Trinkhaus, and T. W. Holiday claim that human brain size, relative to body size, changed little over the last 1.8 million years; see "Body Mass and Encephalization in Pleistocene *Homo*," *Nature* 387 (1997): 173–76. This, I'm sure, is a minority opinion.

34. G. S. Krantz, "*Homo erectus* Brain Sizes by Subspecies," *Human Evolution* 10 (1995): 107–17.

35. Sue Armstrong, "Labour of Death," *New Scientist* 1710 (1990): 50–55.

36. Peter Ellison, *On Fertile Ground* (Cambridge, Mass.: Harvard University Press, 2001), p. 53.

37. Craig Stanford derides the nutritional importance of meat with remarks such as "tubers and beans make an equally protein-rich diet." See *The Hunting Apes: Meat Eating and the Origins of Human Behavior* (Princeton, N.J.: Princeton University Press, 1999), p. 210. I disagree. Tubers are notoriously poor in protein, while "beans" are sparse and seasonal in most landscapes—and a good percentage are poisonous.

38. M. Sponheimer and J. Lee-Thorp, "Isotopic Evidence for the Diet of An Early Hominid, *Australopithecus africanus*," *Science* 283 (1999): 368–70.

39. G. E. McClearn et al., "Substantial Genetic Influence on Cognitive Abilities in Twins Eighty or More Years Old," *Science* 276 (1997): 1560–63.

40. Similarly, the heritability of musical pitch recognition has recently been estimated as 71 to 80 percent; see D. Drayna et al., "Genetic Correlates of Musical Pitch Recognition in Humans," *Science* 291 (2001): 1969–72.

41. M. L. McKinney, "The Juvenilized Ape Myth—Our 'Overdeveloped' Brain," *Bioscience* 48 (1998): 109–16.

42. John C. Eccles, *The Human Mystery* (London: Routledge and Kegan, 1979), p. 234. Steven Mithen also sees a fundamental difference between the minds of chimpanzees and humans. In *The History of the Mind* (London: Thames and Hudson, 1996), he constructs a theory claiming that the human mind became fully integrated only in modern humans. I believe that gradual change over millions of years is the more likely scenario. See Patricia Churchland, "Self-Representation in Nervous Systems," *Science* 296 (2002): 308–10.

43. Terrence Deacon, *The Symbolic Species: The Co-evolution of Language and the Brain* (New York: W. W. Norton, 1997), p. 339.

44. Philosopher Mary Midgley provides an excellent overview of the long tradition of separating mind and body in Western thought: see *The Ethical Primate, Humans, Freedom, and Morality* (London: Routledge, 1994).

45. Sue Savage-Rumbaugh and R. Lewin, *Kanzi: The Ape at the Brink of the Human Mind* (New York: John Wiley, 1994), p. 135.

46. Roger Fouts, *Next of Kin: What Chimpanzees Have Taught Me about Who We Are* (New York: Morrow, 1997).

47. H. L. Miles, "How Can I Tell a Lie? Apes, Language, and the Problem of Deception," in *Deception: Perspectives on Human and Non-human Deceit*, ed. R. W. Mitchell and N. S. Thompson (Albany: State University Press of New York: 1986), pp. 245–66.

48. Ernst Mayr, *Animal Species and Evolution* (Cambridge, Mass.: Harvard University Press, 1963), p. 650.

49. L. Radinsky, "Primate Brain Evolution," *American Scientist* 63 (1975): 656–63.

## 8. BUILDING THE WORLD'S BEST BRAIN

1. S. Semaw et al., "2.5-million-year-old Stone Tools from Gona, Ethiopia," *Nature* 385 (1997): 333–36.

2. In almost any other animal phylogeny, the two genera would probably be considered as one. The upright posture is a unique feature with which to distinguish a more inclusive genus *Homo*.

3. For a fine review of the importance of the hand and how it has shaped

our evolution, see Frank R. Wilson, *The Hand: How Its Use Shapes the Brain, Language, and Human Culture* (New York: Pantheon Press, 1998).

4. W. R. Leonard and M. L. Robertson, "Rethinking the Energetics of Bipedality," *Current Anthropology* 38 (1997): 304–309.

5. Katherine Milton, "A Hypothesis to Explain the Role of Meat-Eating in Human Evolution," *Evolutionary Anthropology* 8 (1999): 11–20.

6. Alan Walker and Richard Leakey, *The Nariokotome Homo erectus Skeleton* (Cambridge, Mass.: Harvard University Press, 1993).

7. A. Gibbons, "Ancient Island Tools Suggest *Homo erectus* Was a Seafarer," *Science* 279 (1998): 1635–37.

8. For example, a portion of an upper jaw from Spain, dated at about 780,000 years old, has recently been named *Homo antecesor*, together with the claim that this represents a "species" ancestral to Neanderthals; see J. M. Bermúdez de Castro et al., "A Hominid from the Lower Pleistocene of Atapuerca, Spain; Possible Ancestors to Neanderthals and Modern Humans," *Science* 276 (1997): 1392–95.

9. Other many-species scenarios are listed under note 27, below. However, these multiple-species views should be contrasted with the view of Milford Wolpoff and Rachel Caspari, who discuss human variability in some detail; see *Race and Human Evolution: A Fatal Attraction* (New York: Simon and Schuster, 1997).

10. Paul Mellars, *The Neanderthal Legacy: An Archaeological Perspective* (Princeton, N.J.: Princeton University Press, 1995); James Shreeve, *The Neanderthal Enigma: Solving the Mystery of Modern Human Origins* (New York: Willam Morrow, 1995).

11. T. W. Holiday, "Postcranial Evidence of Cold Adaptation in European Neandertals," *American Journal of Physical Anthropology* 104 (1997): 245–58.

12. O. Bar-Yosef and B. Vandermeersch, "Modern Humans in the Levant," *Scientific American* 268, no. 4 (April 1993): 94–100.

13. It has also been claimed that the pollen used as evidence for a floral burial (other plant parts were not preserved) had been brought into the burial site by rodents. This seems a stretch to me.

14. Perhaps, as James Shreeve suggests, "Neanderthals lacked the power to innovate because, to them, innovation was anathema!" See *The Neanderthal Enigma*, p. 339.

15. Richard G. Klein and K. Cruz-Uribe, "Exploitation of Large Bovids and Seals at Middle and Later Stone Age Sites in South Africa," *Journal of Human Evolution* 31 (1996): 315–34.

16. F. J. Ayala, "Genes and Origins: The Story of Modern Humans," *Journal of Molecular Evolution* 41 (1995): 683–88.

17. M. F. Hammer et al., "Out of Africa and Back Again: Nested Cladistic Analysis of Human Y Chromosome Variation," *Molecular Biology and Evolution* 15 (1998): 427–41.

18. C. Loring Brace, *The Stages of Human Evolution*, 5th ed. (New York: Simon and Schuster, 1995), p. 224.

19. Recent dating of *erectus*-like skull caps in Indonesia at 25,000 to 40,000 years ago came as a surprise to many, but also substantiated earlier interpretations; see C. C. Swisher III et al., "Latest *Homo erectus* of Java: Potential Contemporaneity with *Homo sapiens* in Southeast Asia," *Science* 270 (1996): 1870–74.

20. Ian Tattersall, "Species Recognition in Human Paleontology," *Journal of Human Evolution* 15 (1986): 165–75.

21. Stephen Stanley, *Children of the Ice Age: How a Global Catastrophe Allowed Humans to Evolve* (New York: Harmony Books, 1996).

22. Stephen J. Gould and Niles Eldredge claimed that evolutionary trajectories were marked by long periods of stasis and short bursts of change, providing an important counterpoint to the traditional emphasis on continous change over time; see S. J. Gould and N. Eldredge, "Punctuated Equilibria—The Tempo and Mode of Evolution Reconsidered," *Paleobiology* 3 (1977): 115–51. Today, the question is not whether evolutionary histories follow a gradual trajectory or one marked by long static periods (both occur) but the relative importance of the alternative scenarios.

23. Susan Cachel, "The Theory of Punctuated Equilibria and Evolutionary Anthropology," in *The Dynamics of Evolution: The Punctuated Equilibrium Debate in the Natural and Social Sciences*, ed. Albert Somit and S. A. Peterson (Ithaca, N.Y.: Cornell University Press, 1989), pp. 187–220.

24. Some scholars have claimed that *sapiens* and Neanderthal people were unlikely to have sexual relations. From the behavior of people today, this seems unlikely.

25. Alan Templeton, "Human Race: An Evolutionary Perspective," Lecture at Field Museum, Chicago, 29 July 1998.

26. Wolpoff and Caspari, *Race and Human Evolution.*

27. Books and articles that espouse multiple speciation events in our lineage over the last two million years include the following: Robert Foley, *Humans before Humanity* (New York: Oxford University Press, 1995); Donald Johanson and B. Edgar, *From Lucy to Language* (New York: Simon and Schuster, 1996); Christopher Stringer and Robin McKie, *African Exodus: The Origins of Modern Humanity* (New York: Henry, 1996); Ian Tattersall, "Once We Were Not Alone," *Scientific American* 282, no. 1 (January 2000): 56–62.

28. Stephen Stanley, *Children of the Ice Age: How a Global Catastrophe Allowed Humans to Evolve* (New York: Harmony Books, 1996).

29. William H. Calvin, *The Ascent of Mind: Ice Age Climates and the Evolution of Intelligence* (New York: Bantam Books, 1991).

30. Richard Potts, *Humanity's Descent: The Consequences of Ecological Instability* (New York: William Morrow, 1996).

31. Paul Colinvaux and P. De Oliveira, "Amazon Plant Diversity and Climate through the Cenozoic," *Palaeogeography, Palaeoclimatology and Palaeoecology* 166 (2001): 51–63.

32. D. A. Livingstone, "Later Quaternary Climatic Change in Africa," *Annual Review of Ecology and Systematics* 6 (1975): 249–80.

33. During the 1960s, the Imperial College of Agricultural and Mechanical Arts was supported by the United States Agency for International Development and administered through Oklahoma State University. A significant number of our graduates went on to acquire doctoral degrees in Europe and the United States.

34. Some of these lowland thornbush baboon populations were studied by Swiss zoologist Hans Kummer.

35. Derek Roe, "Summary and Overview," in *Plio-Pleistocene Archaeology: Koobi Fora Research Project*, Vol. 5, ed. Glynn Isaac (New York: Oxford University Press, 1997), p. 562.

36. This point of view stands in strong contrast to "the climate did it" stance found in other recent books. Compare Noel T. Boaz, *Eco Homo: How the Human Being Emerged from the Cataclysmic History of the Earth* (New York: Basic Books, 1997); Calvin, *The Ascent of Mind*; Potts, *Humanity's Descent*; and Stanley, *Children of the Ice Age*.

37. Katherine Milton, "Diet and Primate Evolution," *Scientific American* 269, no. 2 (August 1993): 86–93; K. Milton, "A Hypothesis to Explain the Role of Meat-Eating in Human Evolution," *Evolutionary Anthropology* 8 (1999): 11–20.

38. Dean Falk, "Brain Evolution in *Homo*: The 'Radiator' Theory," *Behavioral and Brain Science* 13 (1990): 331–81; Dean Falk, *Brain Dance: New Discoveries about Human Origins and Brain Evolution* (New York: Henry Holt, 1992).

39. In addition, Ofer Bar-Yosef and Steven Kuhn argue that there is no justification for linking the more modern "laminar technologies" to any particular aspect of recent behavioral or anatomical evolution; see "The Big Deal about Blades: Laminar Technologies and Human Evolution," *American Anthropology* 101 (1999): 322–38.

40. Glynn Isaac, "The Food-Sharing Behavior of Protohuman Hominids," *Scientific American* 238, no. 4 (April 1978): 90–106.

41. William H. Calvin, *The River That Flows Uphill* (New York: MacMillan, 1986).

42. Nancy M. Tanner, *On Becoming Human* (Cambridge, England: Cambridge University Press, 1981).

43. Robin Dunbar, *Grooming, Gossip, and the Evolution of Language* (Cambridge, Mass.: Harvard University Press, 1996); R. Dunbar, "The Social Brain Hypothesis," *Evolutionary Anthropology* 6 (1998): 178–90.

44. Terrence Deacon, *The Symbolic Species: The Co-evolution of Language and the Brain* (New York: W. W. Norton, 1997).

45. Frans de Waal, *Chimpanzee Politics: Power and Sex among Apes*, rev. ed. (Baltimore: Johns Hopkins University Press, 1998).

46. John L. Locke, "Rank and Relationship in the Evolution of Spoken Language," *Journal of the Royal Anthropological Institute* (N.S.) 7 (2001): 37–50.

47. Matt Ridley, *The Red Queen: Sex and the Evolution of Human Nature* (Ontario, Canada: Maxwell MacMillan, 1993).

48. M. F. Ashley Montagu, "The New Litany of 'Innate Depravity,' or Original Sin Revisited," in *Man and Aggression*, ed. M. F. Ashley Montagu (New York: Oxford University, 1968), p. 11. For a very different view, see Jerome H. Barkow et al., eds., *The Adapted Mind: Evolutionary Psychology and the Generation of Culture* (New York: Oxford University Press, 1992).

49. Intellectually, this is a stance perfectly compatible with a human origin in the Garden of Eden.

50. Robert Ardrey, *The Territorial Imperative: A Personal Inquiry into the Animal Origins of Property and Nations* (New York: Athenium, 1966). See also chapter 2 in John Keegan's *A History of Warfare* (New York: Alfred A. Knopf, 1993).

51. Michael Ghiglieri, *The Dark Side of Man; Tracing the Origins of Male Violence* (Reading, Mass.: Perseus Books, 1999).

52. See for example, William Wright's *Born That Way: Genes, Behavior, Personality* (New York: Alfred A. Knopf, 1998).

53. Ralph L. Beals, Harry Hoijer, and Alan R. Beals, *An Introduction to Anthropology*, 5th ed. (New York: Macmillan, 1977), p. 121.

54. Roger Pitt, "Warfare and Hominid Brain Evolution," *Journal of Theoretical Biology* 72 (1978): 551–75.

55. Steve Jones, Robert Martin, and David Pilbeam, *The Cambridge Encyclopedia of Human Evolution* (Cambridge, England: Cambridge University Press, 1992), p. 404.

56. Richard Leakey and Roger Lewin, *The Sixth Extinction: Patterns of Life and the Future of Mankind* (New York: Doubleday, 1995).

57. Jane Goodall, "Gombe Chimpanzee Politics," in *Primate Politics*, ed. Glendon Schubert and Roger D. Masters (Carbondale, Ill.: Southern Illinois University Press, 1991), pp. 105–37.

58. R. Dyson-Hudson, "Intertribal Hostilities," *Science* 207 (1980): 170–71.

59. For a more comprehensive discussion of intergroup conflict during our long preagricultural history see A. Gat, "The Human Motivational Complex: Evolutionary Theory and the Causes of Hunter-Gatherer Fighting," *Anthropological Quarterly* 73 (2000): 20–34, 74–88.

60. Charles Darwin, *The Origin of Species by Means of Natural Selection, or the Preservation of Favored Races in the Struggle for Life*, 6th ed. (New York: D. Appleton, 1895), pp. 128–32.

61. Richard D. Alexander, "The Evolution of Social Behavior," *Annual Review of Ecology and Systematics* 5 (1974): 325–83; Richard D. Alexander, *The Biology of Moral Systems* (Hawthorne, N.Y.: Aldine De Gruyter, 1987).

62. Paul Bingham argues that our novel "remote-killing capability . . . inevitably led to an unprecedented social revolution: large-scale kinship-independent conspecific cooperation." But kinship, however diffuse, has never relinquished its hold on the human psyche. See Bingham's "Human Evolution and Human History: A Complete Theory," *Evolutionary Anthropology* 9 (2000): 248–58.

63. Mary Midgley, *The Ethical Primate, Humans, Freedom and Morality* (London, Routledge, 1994), p. 119.

64. Frans de Waal, *Good Natured: The Origin of Right and Wrong in Humans and Other Animals* (Cambridge, Mass.: Harvard University Press, 1996).

65. Richard Wrangham and Dale Peterson, *Demonic Males: Apes and the Origins of Human Violence* (Boston: Houghton Mifflin, 1996).

66. Lyall Watson, *Dark Nature: A Natural History of Evil* (New York: HarperCollins, 1995).

67. Jane Goodall, "Gombe Chimpanzee Politics," in *Primate Politics*, ed. G. Schubert and R. D. Masters (Carbondale: Southern Illinois University Press, 1991), pp. 105–37.

68. Irenaeus Eibl-Eibesfeldt, *The Biology of Peace and War* (New York: Viking Press, 1979), p. 122.

69. Geerat J. Vermeij, *Evolution and Escalation: An Ecological History of Life* (Princeton, N.J.: Princeton University Press, 1987), p. 21.

70. Jacques Monod, *Chance and Necessity: An Essay on the Natural Philosophy of Modern Biology* (New York: Alfred A. Knopf, 1971), p.161.

71. Edward O. Wilson, *On Human Nature* (Cambridge, Mass.: Harvard University Press, 1978), p. 116.

72. Though popular for a while, the idea that warfare was insignificant before the creation of modern city-states is no longer tenable; see Lawrence H. Keeley, *War before Civilization* (New York: Oxford University Press, 1996).

73. Y. Fernández-Jalvo et al., "Evidence of Early Cannibalism," *Science* 271 (1996): 277–78. These finds are similar to earlier work in southern France; see P. Villa et al., "Cannibalism in the Neolithic," *Science* (1986): 431–37. For a short review, see Tim D. White, "Once Were Cannibals," *Scientific American* 285, no. 2 (August 2001): 58–65.

74. Barbara Ehrenreich, *Blood Rites: Origins and History of the Passions of War* (New York: Henry Holt and Co., 1997).

75. See the introduction to John Keegan, *A History of Warfare* (New York: Alfred A. Knopf, 1993).

76. As William Howells phrases it, we appear to be a species "whose members in varying degrees tend to find their emotional comfort and security in groups that reinforce that sense by antagonism to other groups." *Getting Here: The Story of Human Evolution* (Washington, D.C.: Compass Press, 1993), p. 230.

77. Elliott Sober and David S. Wilson, *Unto Others: The Evolution and Psychology of Unselfish Behavior* (Cambridge, Mass.: Harvard University Press, 1998).

78. J. N. Davis and M. Daly, "Evolutionary Theory and the Human Family," *Quarterly Review of Biology* 72 (1997): 407–35.

79. Ardrey, *The Territorial Imperative*, p. 350.

80. Richard Dawkins and J. R. Krebs, "Arms Races between and within Species," *Proceedings of the Royal Society, London* B 205 (1979): 489–511.

81. David Premack, *"Gavagai!" or the Future History of the Animal Language Controversy* (Cambridge, Mass.: MIT Press, 1986), p. 133.

82. Will Durant and Ariel Durant, *The Lessons of History* (New York: Simon and Schuster, 1968), p. 19.

83. Richard D. Alexander, *The Biology of Moral Systems* (Hawthorne, N.Y.: Aldine De Gruyter, 1987); R. D. Alexander, "Evolution of the Human Psyche," in *The Human Revolution*, ed. Peter Mellars and Bruce Stringer (Edinburgh, Scotland: Edinburgh University Press, 1987), pp. 455–513. See also R. Wrangham, "Evolution of Coalitionary Killing," *Yearbook of Physical Anthropology* 42 (1999): 1–30.

## 9. AN APPETITE FOR ENERGY: REVOLUTONS IN LIVING

1. This is what some are calling a Lamarckian form of cultural evolution. Lamarck had thought that characteristics acquired during one's lifetime could affect inheritance; they can't. But cultural innovations can be transmitted in the way Lamarck had suggested.

2. T. Goebel, "Pleistocene Human Colonization of Siberia and Peopling of the Americas: An Ecological Approach," *Evolutionary Anthropology* 8 (1999): 208–27.

3. Paul S. Martin and Richard Klein, eds. *Quaternary Extinctions: A Prehistoric Revolution*, (Tucson: University of Arizona Press, 1989). See also J. Alroy, "A Multispecies Overkill Simulation of the End-Pleistocene Megafaunal Mass Extinction," *Science* 292 (2001): 1893–96.

4. Peter D. Ward, *The Call of Distant Mammoths: Why the Ice Age Mammals Disappeared* (New York: Copernicus, 1997), p.220.

5. G. F. Miller et al., "Pleistocene Extinction of *Genyornis newtoni*: Human Impact on Australian Megafauna," *Science* 283 (1999): 205–208.

6. M. C. Fountain, "The Achilles' Heel of Females in Sports," *Chicago Tribune* (6 May 2001): sec. 13, p. 3.

7. For a well-referenced review of the evolutionary basis of human gender differences, see Alison Jolly's *Lucy's Legacy: Sex and Intelligence in Human Evolution* (Cambridge, Mass.: Harvard University Press, 1999). George Williams discusses some of the selective conflicts underlying human reproductive biology in chapter 6 of *The Pony Fish's Glow and Other Clues to Plan and Purpose in Nature* (New York: Basic Books, 1997).

8. Robert Pool, *Eve's Rib: Searching for the Biological Roots of Sex Differences* (New York: Crown Press, 1994).

9. I. Silverman and M. Eals, "Sex Differences in Spatial Abilities: Evolutionary Theory and Data," in *The Adapted Mind: Evolutionary Psychology and the Generation of Culture*, ed. Jerome Barkow et al. (New York: Oxford University Press, 1992), pp. 533–49.

10. Compare also G. Mitchell, "Human Sex Differences. A Primatologist's Perspective," (New York: Van Nostrand, 1981).

11. For a discussion of hormonal effects and other aspects of male-female differences, see Deborah Blum's *Sex on the Brain: The Biological Differences between Men and Women* (New York: Viking, 1997).

12. Over their many years of childhood, my daughters never got into wrestling or punching matches; they never "shot" anybody with broomsticks or other surrogate devices. Despite their father's enthusiastic promotion, they showed no interest in trucks. Sad to say, some of these attributes seem to be inborn.

13. For boys, greater variance should produce a "bell curve" (normal statistical distribution pattern) of test scores in which the center is not as high, and with thicker "tails" at either end for the not-so-smart and the gifted. If their variance is less, the bell-shaped curve for girls should be higher in the middle and with thinner tails. But note that the average or median, at the midpoint on the curve, may be identical for the two groups.

14. This argument explains the prevalence of madness, genius, schizophrenia, and similar syndromes as a "side effect" (in part) of the general value of variability in *all traits* for long-term population survival.

15. Zhao Zhijun, "The Middle Yangtze Region in China Is One Place Where Rice Was Domesticated; Phytolith Evidence from the Diaotonghuan Cave, Northern Jiangxi," *Antiquity* 72 (1998): 885–97.

16. B. D. Smith, "The Initial Domestication of *Cucurbita pepo* in the Americas Ten Thousand Years Ago," *Science* 276 (1997): 932–34.

17. Andrew Sherrat believes that pulsed climatic change may have been a key force in the origin of agriculture in the Middle East; see "Climatic Cycles and Behavioral Revolutions: Emergence of Modern Humans and the Beginnings of Farming," *Antiquity* 71 (1997): 271–87.

18. For a review of the origins of agriculture and our major domesticates, see Jack Harlan's *The Living Fields: Our Agricultural Heritage* (Cambridge, England: Cambridge University Press, 1995).

19. Luigi Cavalli-Sforza and F. Cavalli-Sforza, *The Great Human Diasporas* (Reading, Mass.: Addison-Wesley, 1995).

20. Joel Cohen estimates that the human population ranged somewhere between 2 and 20 million before the agricultural revolution and expanded to between 170 and 330 million by the beginning of the Christian era; see *How Many People Can the Earth Support?* (New York: W. W. Norton, 1995), p. 77.

21. For reviews of plant and animal domestication see Bruce Smith, *The Emergence of Agriculture* (New York: Scientific American Library, 1995), and Jared Diamond's "Evolution, Consequences, and the Future of Plant and Animal Domestication," *Nature* 418 (2002): 700–707.

22. C. S. Larsen, "Biological Changes in Human Populations with Agriculture," *Annual Review of Anthropology* 24 (1996): 185–213.

23. Roy Porter, *The Greatest Benefit to Mankind, A Medical History of Humanity* (New York: W. W. Norton, 1998), p. 18.

24. Dr. Jorge León, speaking at a conference on plant introduction in Zamorano, Honduras, in 1968.

25. Jared Diamond, *Guns, Germs and Steel: The Fates of Human Societies* (New York: W. W. Norton, 1997).

26. Juliette Clutton-Brock, *Horse Power: A History of the Horse and Donkey in Human Societies* (Cambridge, Mass.: Harvard University Press, 1992).

27. Charles Stanish, "Origins of the State in the Titicaca Basin of Southern Peru," Lecture at Field Museum, Chicago, 1997.

28. Donald Cardwell, *The Norton History of Technology* (New York: W. W. Norton, 1995), p. 17.

29. Until a few hundred years ago, forges could not reach temperatures higher than about 1,200°C.

30. Stephen Sass, *The Substance of Civilization: Material and Human History from the Stone Age to the Age of Silicon* (New York: Arcade Publishing, 1998), p. 91.

31. W. Rostoker and B. Bronson, *Pre-Industrial Iron: Its Technology and Ethnology* (Philadelphia, Archaeomaterials Monograph No. 1, 1990).

32. Jean Gimpel, *The Medieval Machine: The Industrial Revolution of the Middle Ages* (New York: Holt, Rinehart and Winston, 1976); and James McClellan III and Harold Dorn, *Science and Technology in World History* (Baltimore, Johns Hopkins University Press, 1999).

33. Hubert H. Lamb, *Climate, History, and the Modern World*, 2d ed. (London: Routledge, 1995).

34. Useful references for this period are: Robert Bartlett, *The Making of Europe: Conquest, Colonization, and Cultural Change: 950–1350* (Princeton, N.J.: Princeton University Press, 1993); Frances and Joseph Gies, *Cathedral, Forge, and Waterwheel: Technology and Invention in the Middle Ages* (New York: HarperCollins, 1994); and Henri Pirenne, *Medieval Cities: Their Origins and the Revival of Trade* (Princeton, N.J.: Princeton University Press, 1925).

35. David S. Landes, *The Wealth and Poverty of Nations* (New York: W. W. Norton, 1998), p. 6.

36. For an overview, see McClellan and Dorn, *Science and Technology in World History*.

37. Robert McCormick Adams, *Paths of Fire: An Anthroplogist's Inquiry into Western Technology* (Princeton, N.J.: Princeton University Press, 1996), p. 69.

38. J. R. McNeill, *Something New under the Sun: An Environmental History of the Twentieth Century* (New York: W. W. Norton, 2000), p. 14.

39. W. H. McNeill, "American Food Crops in the Old World," in *Seeds of Change*, ed. Herman J. Viola and Carolyn Margolis (Washington, D.C.: Smithsonian Institution Press, 1991), p.52.

40. Lindsey Grant, *Juggernaut: Growth on a Finite Planet* (Santa Ana, Calif.: Seven Locks Press, 1996), p. 23. See also Vaclav Smil, *Enriching the Earth: Frits Haber, Carl Bosch, and the Transformation of World Food Production* (Cambridge, MIT Press, 2000).

41. Peter Huber claims that "With markets in command, scarcity is always giving way to abundance . . . we keep getting richer." (*Hard Green: Saving the Environment from the Environmentalists: A Conservative Manifesto* [New York: Basic Books, 1999], p. 14). I disagree; we're getting richer because we are using more and more energy to devour more resources.

# 10. THE ORIGINS OF WESTERN SCIENCE

1. Useful references are Robert Bartlett, *The Making of Europe: Conquest, Colonization, and Cultural Change: 950–1350* (Princeton, N.J.: Princeton University Press, 1993); Will Durant, *The Age of Faith: The Story of Civilization: Part IV* (New York: Simon and Schuster, 1950).

2. G. Warren Hollister, *Medieval Europe: A Short History*, 6th ed. McGraw-Hill, 1990), p. 266.

3. Otto von Simson, *The Gothic Cathedral*, 2d ed. (New York: Pantheon Books, 1962).

4. David Lindberg, *The Beginnings of Western Science: The European Scientific Tradition in Philosophical, Religious, and Institutional Context: 600 B.C. to A.D. 1450* (Chicago, University of Chicago Press, 1992), p. 200.

5. Richard Tarnas, *The Passion of the Western Mind: Understanding the Ideas That Have Shaped Our World View* (New York: Balantine Books, 1991), p. 176.

6. Norman F. Cantor, *The Civilization of the Middle Ages* (New York: Harper Perennial, 1993), p. 320.

7. Edward Grant, *The Foundation of Modern Science in the Middle Ages* (Cambridge, England: Cambridge University Press, 1996), p. 21.

8. Ernest Gellner, *Plough, Sword and Book: The Structure of Human History* (Chicago, University of Chicago Press, 1988), p. 82.

9. Nancy G. Sirasi, *Medieval and Early Renaissance Medicine* (Chicago, University of Chicago Press, 1990); Andrea Carlino, *Books of the Body: Anatomical Ritual and Renaissance Learning* (Chicago, University of Chicago, 1999).

10. Christians may have come to view postmortem dissections with less repulsion than other faiths because they separated the concept of the immortal soul from one's transient physical abode according to Daniel Boorstin, *The Discoverers* (New York: Harry N. Abrams, 1991), p. 537.

11. Roy Porter, *The Greatest Benefit to Mankind: A Medical History of Humanity* (New York: W. W. Norton, 1998), p. 132.

12. The Chinese had sent a huge expedition into the Indian Ocean earlier, but it had no well-defined purpose, and its great expense discouraged further efforts.

13. Good general references are Boorstin, *The Discoverers* and David Divine, *The Opening of the World, The Great Age of Maritime Exploration* (New York: G. P. Putnam, 1973).

14. Bartlett, *The Making of Europe*; John Hale, *Age of Exploration: Great Ages of Man* (New York: Time-Life Books, 1974).

15. John Hale, "Geographical Horizons and Mental Horizons," in *The Age of the Renaissance*, ed. Denys Hay (New York: McGraw Hill, 1967), p. 338.

16. M. Clapham, "Printing," in *A History of Technology, Vol. 3. From the Renaissance to the Industrial Revolution*, ed. Charles Singer et al. (Oxford, England: Oxford University Press, 1957), pp. 377–416.

17. Thomas Goldstein, *Dawn of Modern Science: From the Arabs to Leonardo da Vinci* (Boston: Houghton Mifflin, 1980), p. 171.

18. Richard Powers, "Eyes Wide Open: When an Obscure Arab Scientist Solved the Riddle of Light, the Universe No Longer Belonged to God," *New York Times Magazine* (18 April 1999): 83.

19. Alan Cromer, *Uncommon Sense: The Heretical Nature of Science* (New York: Oxford University Press, 1993), p. 70.

20. Ibid., p. 99.

21. Some readers may object that this is a narrowly Eurocentric view. However, no other society became so supportive of public enterprise, both intellectual and mercantile, relatively free of either religious or political bureaucracy, for so long. See also Nathan Rosenberg and L. E. Birdzelk, *How the West Grew Rich: The Economic Transformation of the Industrial World* (New York: Basic Books, 1986).

22. James McClellan III and Harold Dorn, *Science and Technology in World History* (Baltimore: Johns Hopkins University Press, 1999).

23. William H. McNeill, *The Rise of the West: A History of the Human Community* (Chicago, University of Chicago Press, 1963), p. 545.

24. Harold J. Berman, *Law and Revolution: The Formation of the Western Legal Tradition* (Cambridge, Mass.: Harvard University Press, 1983). Norman Cantor also sees the investiture dispute as having created a major historical revolution. However, for Cantor, it was the removal of religious authority from the claims of royal and political leadership that marked "the creation of a new world order," *The Civilization of the Middle Ages*, p. 258.

25. Berman, *Law and Revolution*, p. 529.

26. David Fromkin, *The Way of the World: From the Dawn of Civilizations to the Eve of the Twenty-first Century* (New York: Alfred A. Knopf, 1998), p. 75.

27. Durant, *The Age of Faith*, pp. 916–17.

28. Toby E. Huff, *The Rise of Early Modern Science: Islam, China, and the West* (Cambridge, England: Cambridge University Press, 1993), p. 212.

29. Lindberg, *The Beginnings of Western Science*, p. 208.

30. Jacques Barzun, *From Dawn to Decadence: Five Hundred Years of Western Cultural Life: 1500 to the Present* (New York: HarperCollins, 2000), p. 230.

31. Marcia Colish, *Medieval Foundation of the Western Intellectual Tradition: 400—1400* (New Haven, Conn.: Yale University Press, 1997), p.272.

32. Grant, *The Foundation of Modern Science in the Middle Ages*, pp. 193, 206.

33. Tarnas, *The Passion of the Western Mind*, p. 231.

34. See also Jean Gimpel, *The Medieval Machine: The Industrial Revolution of the Middle Ages* (New York: Holt, Rinehart and Winston, 1976).

35. Brian Swimme and T. Berry, *The Universe Story* (San Francisco: Harper, 1992), p. 226.

## 11. PERFECT PLANET, BUT HOW UNUSUAL AN ODYSSEY?

1. Thomas Berry, "The Story and the Dream: The Next Stage of the Evolutionary Epic," Lecture in conference: *The Epic of Evolution*, Field Museum, Chicago, 14 November 1997.

2. Isaac Asimov, *Extraterrestrial Civilizations* (New York: Crown, 1979), p. 197.

3. For a very different view ("It is quite clearly nonsense to suggest that we will never get to the stars. . . .") see Michael White, *Life Out There: The Truth of—and Search for—Extraterrestrial Life*, American edition, (Hopewell, N.J.: Ecco Press, 1998), p. 163.

4. Adrian Berry, *The Next Five Hundred Years: Life in the Coming Millenium* (New York: W. H. Freeman, 1995), p. 197.

5. Michael H. Hart, "An Explanation for the Absence of Extraterrestrials on Earth," *Quarterly Journal of the Royal Astronomical Society* 16 (1975): 128–35; Frank J. Tipler, "Extraterrestrial Intelligent Beings Do Not Exist," *Quarterly Journal of the Royal Astronomical Society* 21 (1980): 267–81. See also G. D. Brin, "The Great Silence: The Controversy Concerning Extraterrestrial Intelligent Life," *Quarterly Journal of the Royal Astronomical Society* 24 (1983): 283–309.

6. Frank Drake, quoted in Steven J. Dick, *Life on Other Worlds: The Twentieth-Century Extraterrestrial Life Debate* (Cambridge, England: Cambridge University Press, 1998) p. 220. See also David Koerner and Simon LeVay, *Here Be Dragons: The Scientific Quest for Extraterrestrial Life* (New York: Oxford University Press, 2000), pp. 171–88.

7. Recent popular books discussing the probability of extraterrestrial intelligence that I found especially informative are Andrew Clark and David Clark, *Aliens: Can We Make Contact with Extraterrestrial Intelligence?* (New York: Fromm International, 1999); David Darling, *The Extraterrestrial Encyclopedia*

(New York: Three Rivers Press, 2000); Frank Drake and Dava Sobel, *Is Anyone Out There? The Scientific Search for Extraterrestrial Intelligence* (New York: Delacorte Press, 1992); Koerner and LeVay, *Here Be Dragons*; Walter Sullivan, *We Are Not Alone: The Continuing Search for Extraterrestrial Intelligence*, rev. ed. (New York: Dutton, 1993); and Peter Ward and Donald Brownlee, *Rare Earth: Why Complex Life Is Uncommon in the Universe* (New York: Copernicus Books, 2000). For a short recent review, see T. L. Wilson, "The Search For Extraterrestrial Intelligence," *Nature* 409 (2001): 1110–14.

8. Drake and Sobel, *Is Anyone Out There?* p. 206.

9. Ibid., p. 52.

10. Ken Croswell, *The Alchemy of the Heavens* (New York: Anchor Press, 1995), p. 182.

11. Clark and Clark, *Aliens*, p. 45.

12. These estimates follow Ken Croswell, *The Alchemy of the Heavens* (New York: Anchor Press, 1995), p. 248.

13. Only a few texts examine the many necessary attributes of a planet that can sustain complex life-forms; see John D. Barrow and Frank J. Tipler, *The Anthropic Cosmological Principle* (New York: Oxford University Press, 1986); Clark and Clark, *Aliens*; and especially Ward and Brownlee, *Rare Earth*.

14. A strong case for the importance of water and plate tectonics is made by Ward and Brownlee in *Rare Earth*.

15. Drake and Sobel, *Is Anyone Out There?* p. 209.

16. Armand Delsemme, *Our Cosmic Origins* (Cambridge, England: Cambridge University Press, 1998), p. 239.

17. Amir Aczel, *Probability One: Why There Must Be Intelligent Life in the Universe* (New York: Harcourt Brace, 1998), p. 201.

18. Carl Sagan, *Pale Blue Dot: A Vision of The Human Future in Space* (New York: Random House, 1994), p. 33.

19. Loren Eiseley, *The Immense Journey* (New York: Vintage Books, 1957), p. 77.

20. For an incisive commentary on the unfolding disaster we have initiated and a speculative explanation of its basis, see Greg Morrison's *The Spirit in the Gene: Humanity's Proud Illusion and the Laws of Nature* (Ithaca, N.Y.: Cornell University Press, 1999).

21. Ramesh Thakar, quoted in the *Chicago Tribune* (16 January 2000) sec. 2, p. 7.

22. J. R. McNeill, *Something New under the Sun: An Environmental History of the Twentieth Century* (New York: W. W. Norton, 2000), p. 271.

23. Ibid., p. 359.

24. Albert Gore, *Earth in the Balance: Ecology and the Human Spirit*, Plume edition (New York: Penguin Books, New York: 1993), p. 239.

25. Julian Simon, ed., *The State of Humanity* (Cambridge, Mass.: Blackwell, 1995). For a recent review by those who share Julian Simon's perspective, see *Earth Report 2000: Revisiting the State of the Planet*, ed. Ronald Bailey (New York: McGraw Hill, 2000). Peter Huber presents a "manifesto" for this point of view in *Hard Green: Saving the Environment from the Environmentalists: A Conservative Manifesto* (New York: Basic Books, 1999).

26. J. H. Ausubel and H. D. Langford, eds., *Technological Trajectories and the Human Environment* (Washington, D.C.: National Academy Press, 1997).

27. Ben J. Wattenberg, "The Population Explosion Is Over," *The New York Times Magazine* (23 November 1997): 60–63.

28. Lincoln Allison, *Ecology and Utility: The Philosophical Dilemmas of Planetary Management* (Rutherford, N.J.: Farleigh Dickinson University Press, 1991), pp. 70, 157.

29. Huber, *Hard Green*, p. 109.

30. Paul Kennedy, *Preparing for the Twenty-First Century* (New York Random House, 1993), p. 331.

31. Robert Frank, *Luxury Fever: Why Money Fails to Satisfy in an Era of Excess* (New York: Free Press, 1999). See also Brian Czech, who proposes an alternative value system to halt our profligate ways in *Shoveling Fuel for a Runaway Train: Errant Economists, Shameful Spenders, and a Plan to Stop Them All* (Berkeley: University of California Press, 2000).

32. James McClellan III and Harold Dorn, *Science and Technology in World History* (Baltimore: Johns Hopkins University Press, 1999), p. 373.

33. Garrett Hardin, *The Ostrich Factor: Our Population Myopia* (New York: Oxford University Press, 1999), p. 78.

34. Laurie Garrett, *The Coming Plague: Newly Emerging Diseases in a World Out of Balance* (New York: Farrar, Straus and Giroux, 1994).

35. Andrew and David Clark use SLIME as an acronym for *s*imple *l*ife *i*n a *m*icrobial *e*cosystem in *Aliens*, p. 4. Their book makes a sensible plea for further efforts in SETI research.

36. For a recent review of the "progressive" evolution of complexity and diversity here on planet Earth, see S. B. Carroll, "Chance and Necessity: The Evolution of Morphological Complexity and Diversity," *Nature* 409 (2001): 1102–1109.

37. A. K. Dewdney summarizes this attitude succinctly: "With the peculiar myopia that characterizes western culture, we have come to regard our own development as more or less inevitable, an extension of the Darwinian impera-

tive into the technocultural realm," *Yes, We Have No Electrons: An Eye-Opening Tour through the Twists and Turns of Bad Science* (New York: John Wiley and Sons, 1997), p. 71.

38. William Dembski argues that the question of intelligent design should be reintrodued into the fabric of modern science, but he fails to tell us how to distinguish clearly between "cumulative complexity" (built by mutations, selection, and time) and "contingent complexity" (a product of purposeful design). See *Intelligent Design: The Bridge between Science and Theology* (Downers Grove, Ill: Intervarsity Press, 1999)

39. John D. Barrow and Frank J. Tipler, *The Anthropic Cosmological Principle* (New York: Oxford University Press, 1986); Roger Trigg, *Rationality and Science: Can Science Explain Everything* (Oxford, England: Blackwell, 1993).

40. Paul Davies, *The Mind of God: The Scientific Basis of a Rational World* (New York: Simon and Schuster, 1993), p. 16.

41. Michael J. Denton, *Nature's Destiny: How the Laws of Biology Reveal Purpose in the Universe* (New York: Free Press, 1998).

42. For a single-volume review that makes the unity of scientific knowledge more clearly apparent, see John Gribbin's *Almost Everyone's Guide to Science: The Universe, Life and Everything* (New Haven, Conn.: Yale University Press, 1999).

43. Ralph Waldo Emerson, in *The Oxford Authors: Emerson*, ed. Richard Poirier (New York: Oxford University Press, 1990), p. 241.

44. Lee Smolin, *The Life of the Cosmos* (New York: Oxford University Press, 1997), p. 163.

45. For an ambitious overview, see Eric J. Chaisson, *Cosmic Evolution: The Rise of Complexity in Nature* (Cambridge, Mass.: Harvard University Press, 2001).

# INDEX

clever species

References to figures are in boldface.